Normal Microflora

Normal Microflora

An introduction to microbes inhabiting the
human body

G.W. Tannock

Department of Microbiology,
University of Otago, Dunedin,
New Zealand

CHAPMAN & HALL

London · Glasgow · Weinheim · New York · Tokyo · Melbourne · Madras

Published by Chapman & Hall, 2–6 Boundary Row, London SE1 8HN, UK

Chapman & Hall, 2-6 Boundary Row, London SE1 8HN, UK

Blackie Academic & Professional, Wester Cleddens Road, Bishopbriggs, Glasgow G64 2NZ, UK

Chapman & Hall GmbH, Pappelallee 3, 69469 Weinheim, Germany

Chapman & Hall USA, One Penn Plaza, 41st Floor, New York NY 10119, USA

Chapman & Hall Japan, ITP-Japan, Kyowa Building, 3F, 2-2-1 Hirakawacho, Chiyoda-ku, Tokyo 102, Japan

Chapman & Hall Australia, Thomas Nelson Australia, 102 Dodds Street, South Melbourne, Victoria 3205, Australia

Chapman & Hall India, R. Seshadri, 32 Second Main Road, CIT East, Madras 600 035, India

First edition 1995

© 1995 G.W. Tannock

Typeset in 10/12 pt Paladium by Colset Private Limited, Singapore
Printed in Great Britain by St Edmundsbury Press, Bury St Edmunds, Suffolk

ISBN 0 412 55040 7

Preface

This book is about the microbial species that inhabit our body, and the consequences of the intimate relationships that we share with them. It is intended that the book will provide an introduction to the 'normal microflora' for those studying disciplines within the health sciences, and for those in the food industry where interest in the microbiology of the digestive tract is topical.

Our relationships with the normal microflora provide excellent examples of ecological balances that have evolved between species, and the repercussions that ensue should the balance be tilted in favour of one or another partner. An appreciation of this concept is important for all those who study biological sciences.

The book primarily concerns the microbial inhabitants of the human body. Experimental data that contribute to specific concepts have in some cases been more easily obtained from nonhuman species. Discussion of the normal microflora of species other than humans has therefore been included where appropriate. I thank P. Scott, Graphics Unit, Otago Medical School for the preparation of diagrams. Colleagues who contributed photographs are acknowledged in the appropriate Figure legends. This book is dedicated to my parents, A.I. and F.J. Tannock.

G.W. Tannock,
Department of Microbiology, University of Otago,
Dunedin, New Zealand

1 More than a smell: the complexity of the normal microflora

Most communities encourage regular bathing habits by their members for at least two reasons: to reduce the numbers of skin parasites and disease-producing microbes that can be transmitted from one individual to another; and to remove bodily secretions and the microbes that alter them chemically and contribute to body odour. Anyone exposed to commercial television is aware that soaps, deodorants, foot powders and mouthwashes are designed to combat the social sins of body odour, smelly feet and bad breath. The microbial populations responsible for the activities against which these products are aimed belong to a vast collection of species, mostly bacteria, that inhabit the various body sites, even of healthy humans. This microbial collection, generally referred to as the normal microflora, is acquired soon after birth. It is estimated to consist of about 10^{14} microbial cells, outnumbering the approximately 10^{13} human cells that make up the body. The microbes that constitute the normal microflora, being numerous and metabolically active, therefore have the potential to affect our lives profoundly. Although best known for their contribution to body odour and bad breath, the members of the normal microflora in our daily existence are responsible for more than a smell, as this book will attempt to record.

The first scientific observation of the normal microflora was probably made by Antony van Leeuwenhoek in the 17th century. He used his simple microscopes to examine scrapings collected from dental surfaces and observed microscopic cells of various shapes and sizes. Scientific interest in the activities of the normal microflora has been, however, largely a 20th century phenomenon although concepts and theories expressed early this century doubtless had their basis in observations made in the last decades of the previous century. Knowledge accumulated as the result of observations made by numerous researchers over many years has led to the recognition that the body is always host to a large and varied collection of microbes. Major advances in the study of the normal microflora have been due, however, to technical and conceptual studies made by relatively few individuals. Robert Hungate's investigations of the normal microflora of the

bovine rumen led to the development of techniques by which the obligately anaerobic microbes, which are the numerically dominant members of the microflora of body sites, can be cultivated under laboratory conditions. The development of these methods enabled the complexity of the micro-flora to be investigated, notably by Ed Moore, Lillian Holdeman, Sydney Finegold and their colleagues. Rene Dubos and colleagues at the Rockefeller University developed a microecological approach to the investigation of the normal microflora. This led to the recognition of the importance, for some members of the microflora, of the ability to associate with mam-malian surfaces when inhabiting body sites. This latter work has been pioneered by Dwayne Savage with respect to the gastrointestinal tract and by Ron Gibbons concerning the oral microflora. Knowledge of the influences exerted on mammals by the normal microflora has mostly originated from studies utilizing animals maintained in the absence of microbes (germfree animals) by techniques developed by Reyniers, Miya-kawa, Trexler, Gustafsson and their colleagues.

The 'normal microflora' is the term most commonly used when defining the microbial collection inhabiting the body. Other terms sometimes used are 'normal flora', 'commensals' and 'indigenous microbiota'. Of all of these the strictly correct term is 'indigenous microbiota' since it infers a collection of microscopic creatures that are native to the body. 'Flora' and 'microflora' have an unfortunate botanical connotation derived from the days when bacteria and other microbes were considered to resemble plant cells. Com-mensalism refers to an association between two organisms in which one partner benefits from the relationship while the other obtains neither benefit nor harm. This terminology is inappropriate since, as is recorded in other chapters of this book, the association between a host and the normal microflora is not commensalistic since each partner is influenced markedly by the other. Many scientists would prefer the use of 'indigenous microbiota' since it is the more correct of those listed above. However, 'normal microflora' has been used extensively in the medical literature for many decades, has international recognition, is therefore likely to remain in common usage, and so will be used in this book.

The normal microflora inhabits body sites whose surfaces and cavities are open to the environment. Thus the skin surface, oral cavity, respiratory tract, gastrointestinal tract, and urinary and genital tracts are potential sites of microbial colonization. Host factors minimize the extent of microbial colonization in some of these sites, however, so that the normal micro-flora is limited in health to the skin, oral cavity, upper respiratory tract, gastrointestinal tract, a short region of the urinary tract and the vagina. Unique functions, requiring unique structures or molecules, occur in these body sites. These unique properties provide conditions under which certain species of microbes can flourish while other microbial types, lacking appropriate biochemical or other qualities, cannot. Containing unique molecular, structural and microbial characteristics, each of the body sites

Table 1.1 Ecological definitions

Microbial ecology	The study of the interrelationships between microbes and their environment
Ecosystem	The assemblage of species and the organic and inorganic constituents characterizing a particular site
Habitat	An area having a degree of uniformity in characteristics of ecological significance
Community	The microbes inhabiting a given site
Population	A distinguishable microbial type which can be regarded as a discrete entity within a community

harbouring a normal microflora therefore constitutes an ecosystem. The physical or chemical nature of an ecosystem may not be uniform so that microbial distributions may also vary according to these properties. Thus microbial habitats can be recognized within an ecosystem and microbial communities can be recognized and described. Ecological terminology is useful for describing phenomena relating to the normal microflora. Definitions of some of these terms are given in Table 1.1.

As inferred in the preceding paragraph, each body site has its own characteristic microflora. The normal microflora can therefore be divided into the skin (cutaneous) microflora, upper respiratory tract microflora, oral cavity microflora and so on.

1.1 SKIN

The skin provides a large surface area for microbial colonization, about two square metres in the average human, as well as creases and wrinkles in which microbial colonies can develop. The skin ecosystem is obviously not uniform. Just by examining our own skin, we can see that there are areas of the body surface which are dry (back, underside of forearm), areas which are moist (palm of hand) and some areas that are definitely wet (armpit). The distribution of coarse hair also varies according to body site, as does the distribution of sweat glands. Mary Marples has likened this variation in skin characteristics to the varying topography of the earth: 'the desert of the forearm', 'the tropical forest of the armpit', 'the cool woods of the scalp'. These dramatically different environments on earth would provide habitats for different species of plants and animals; so too on the skin we find that microbial colonization differs between body sites. Microbes, like other cellular forms, require water for life. Thus we find that microbes are most numerous on the moist areas of the skin. A dry area, such as the surface of the back, harbours only a few hundred bacteria per square centimetre; a moist area, such as the armpit, harbours several million bacteria per square centimetre (Table 1.2).

Table 1.2 Numbers of aerobic bacteria inhabiting the skin surface

Skin site	Count per square centimetre	
	Contact method	Scrub method
Forehead	348	2.0×10^5
Armpit	106	2.4×10^6
Back	55	3.1×10^2
Forearm	41	1.0×10^2

Contact method: bacterial colonies are removed from the skin by means of sellotape, agar or velvet pressed on to the area to be sampled. The sellotape or velvet is then pressed on to a solid culture medium. Since the accumulations of bacterial cells are only slightly disturbed by these manipulations, counts are usually lower than those obtained using the scrub method.
Scrub method: the area of skin to be sampled is 'scrubbed' with a solution containing a detergent. This is a more efficient method of removing bacteria from the skin surface. Bacterial clumps are disrupted so that counts are higher than those obtained using contact methods. (Source: W.C. Noble and D.A. Somerville, 1974, *Microbiology of Human Skin*, W.B. Saunders, London.)

The availability of water on the skin is governed by two main factors: sweat glands and the sweat that they secrete, and sebaceous glands and the sebum that they secrete. Sebaceous glands are numerous on the forehead and are associated with hair follicles on all parts of the body. The sebum nourishes the hair (Figure 1.1) and spills out of the follicle on to the skin surface, forming an oily mixture over the skin. Some of the inhabitants of the skin can metabolize the lipid constituents of sebum present in the oily layer. This microbial activity makes the surface of some areas of the skin (e.g. the forehead) acidic (pH 5.5), and the oily superficial layer of the skin is sometimes referred to as the 'acid mantle'. The low pH and the presence of fatty acids on the skin surface make it more difficult for many microbial types to survive on the skin, so the acid mantle is a nonspecific defence mechanism of the body against disease-producing organisms. Microbial action on sebum, however, liberates smelly compounds which contribute to 'body odour'. In many societies, therefore, sprays and lotions which contain substances that inhibit microbial replication are applied to the skin. These preparations reduce the amount of microbial activity in the armpit, for example, and act as 'deodorants'.

Sweat glands occur all over the body surface, but they are most densely distributed on the scalp and forehead, and on the soles of the feet and the palms of the hands. The sweat that exudes from pores on the skin provides not only water but also nutrients for microbes. The sodium chloride concentration on the skin surface tends to increase as sweat evaporates and becomes a selective factor for microbial colonization. One of the major microbial populations on the skin surface is composed of salt-tolerant bacteria (the staphylococci). Under laboratory conditions these microbes will grow in media containing 7.5% sodium chloride.

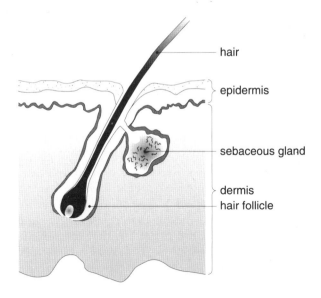

Figure 1.1 General structure of the skin.

Thus far the skin has appeared to be a warm (30–33 °C), wet, pleasant site for microbial colonization. In reality, the skin ecosystem is quite an unstable site for microbes. This instability is due to the structure of the skin and in particular to the nature of the outermost layer of the skin, the stratum corneum. The principal function of the skin is to provide a protective layer to the enclosed body tissues. At the same time, the relatively tough protective layer must still allow flexibility of movement and allow the retention of sensitive contact with the environment. This is achieved by forming a relatively thin but tough epidermis composed of dead and dying cells which contain a protein called keratin. The innermost cells of the epidermis are living and actively replicating. Cells in the middle layers of the epidermis synthesize keratin and contain progressively more and more of this protein as they approach the outermost epidermal layer, stratum corneum. The stratum corneum contains cells that are filled with keratin and are no longer alive. Dead cells on the surface of the stratum corneum are continually, and constantly, peeling off and floating away in the convection currents that surround the body. The loss of superficial cells uncovers new cells underneath which will also be lost eventually. The skin surface, like other epithelia in the body, is always changing as old cells are sloughed off and there is a progressive movement of new cells towards the surface of the stratum corneum. The surface of the human body changes completely within about a two-day period. The epidermis is an example of a keratinized stratified squamous epithelium. The cells that peel off from the skin surface tend to do so in small sheets or clumps. These aggregates of cells are known as squames. As squames leave the body, so too do any microbes that have taken up residence on the skin surface. The microbial burden on the skin is therefore maintained at a much lower level than that encountered in some other body ecosystems. Microbial colonization continues on the skin, however, because as squames leave the body new cells are uncovered which can then be colonized by microbes from the surrounding skin surface.

The rate at which epidermal cells become keratinized can be upset during adolescence by the hormonal changes which occur at that stage of life. Accumulations of keratinized cells can block hair follicles under these circumstances allowing uncontrolled proliferation of bacteria (propionibacteria) trapped within these sites. The blocked follicles, called comedones, only become inflamed causing acne vulgaris if the appropriate pH and concentration of oxygen are present within the follicle. Acne does not occur in all adolescents, which suggests that host factors such as the immunological mechanisms of the body are involved in causing the disease.

Table 1.3 lists the microbes commonly encountered on human skin. Gram-positive bacterial species predominate in the skin ecosystem. At least 12 species of staphylococci can be isolated from human skin (Table 1.3), but the most commonly isolated species are *Staphylococcus epidermidis* (predominating in regions of the upper body) and *S. hominis* (predominating on arms and legs). Staphylococci are Gram-positive cocci that characteristically form aggregates of cells resembling bunches of grapes and are facultatively anaerobic. Micrococci are also Gram-positive. They differ principally from the staphylococci in that micrococci, being obligate aerobes, cannot ferment glucose under anaerobic conditions. *Micrococcus luteus*, a yellow-pigmented species, is the most commonly isolated micrococcus from human skin. It is present mostly on the head, legs and arms.

The Gram-positive bacilli that inhabit the skin belong to a general morphological grouping of bacteria known as the coryneforms. The bacterial cells have a characteristic shape, being short rods with one end of the cell having a slightly greater diameter than the other. Three bacterial genera are represented among the skin coryneforms: *Corynebacterium*, *Propionibacterium* and *Brevibacterium*.

Some of the corynebacteria are obligate aerobes; others are facultative anaerobes. They seem well suited to life on the skin because they produce enzymes that degrade lipids, urea (present in sweat) and proteins. Propionibacteria grow best under anaerobic conditions and obtain their energy by fermenting substrates. The major fermentation product that they produce is propionic acid, which explains the derivation of their generic name. *Propionibacterium acnes* is the most common species of propionibacterium isolated from human subjects past the age of puberty. It is especially prevalent in greasy areas of the skin where there are many sebaceous glands (such as the forehead). The propionibacteria live in the ducts leading from

Table 1.3 Microbes commonly detected on skin

Gram-positive cocci
  Staphylococcus aureus
  S. auricularis
  S. capitis
  S. cohnii
  S. epidermidis
  S. haemolyticus
  S. hominis
  S. saccharolyticus
  S. saprophyticus
  S. simulans
  S. warneri
  S. xylosus
  Micrococcus luteus
  M. lylae
  M. nishinomiyaensis
  M. kristinae
  M. sedentarius
  M. roseus
  M. varians

Gram-positive bacilli
  Corynebacterium jeikeium
  C. urealyticum
  C. minutissimum
  Propionibacterium acnes
  P. avidum
  P. granulosum

  Brevibacterium epidermidis

Gram-negative bacilli
  Acinetobacter calcoaceticus

Yeasts
  Malassezia furfur

Moulds
  Trichophyton mentagrophytes var. *interdigitale*

Mite
  Demodex folliculorum

the sebaceous glands and are therefore well situated to obtain nutrients from the degradation of lipids in sebum. Two other propionibacterial species are present on the skin, though they have more restricted habitats than does *P. acnes*. *Propionibacterium granulosum* is present on the greasy areas at the side of the nose (the alnae nasi) while *P. avidum* prefers moist areas like the armpit.

Brevibacterium epidermidis is common on the skin between the toes. The brevibacteria can convert the amino acid L-methionine to a gas, methane thiol, which is partly responsible for foot odour. Although brevibacteria are proteolytic bacteria, foot odour can be largely controlled by the use of an antifungal powder, suggesting that interactions between brevibacteria and keratin-degrading fungi occur on the skin. Fungi such as *Trichophyton mentagrophytes* var. *interdigitale* degrade keratin, releasing amino acids which may be available to the bacteria. Under constantly moist conditions between the toes keratinolytic fungi often multiply in an uncontrolled manner producing maceration of the skin with resulting discomfort (*tinea pedis*: athlete's foot).

Another fungus encountered in the skin ecosystem is the yeast *Malassezia furfur* (previously known as *Pityrosporum ovale* or *Pityrosporum orbiculare*). The yeast prefers lipid-rich habitats such as the openings of the sebaceous glands of the scalp, external ear and back. The yeast can degrade lipids and waxes to obtain nutrients. It is also associated with alterations to the pigmentation of small areas of the stratum corneum of the trunk and shoulders, a condition known as *pityriasis versicolor*.

Acinetobacter calcoaceticus is an obligately aerobic Gram-negative bacillus whose main habitats are in soil, water and sewage. The bacteria are isolated consistently enough from human skin (about 25% of subjects), however, for them to be considered members of the normal microflora. Acinetobacter usually inhabit the axilla, groin and antecubital fossa (the hollow in front of the elbow). Colonization of the skin by this organism is encouraged by moisture. Hence there is an increased frequency of isolation from the skin during summer due to increased sweating, and from the often wet feet of coalminers.

A small mite (0.3–0.4 mm in length) belonging to the order Acarina is commonly present in hair follicles and the openings of sebaceous glands of the face. The mite, *Demodex folliculorum*, lives so unobtrusively on the skin that its presence is never suspected. Related forms of mite are found on the skin of mammals other than humans, but in these cases the insects are always associated with damage to the ecosystem (mange).

1.2 RESPIRATORY TRACT

The respiratory tract is comprised of the nasal passages, naso- and oropharynx, trachea, bronchi and bronchioles which carry warmed and moistened inhaled air to the lungs and carbon dioxide-laden air away (Figure 1.2). Conditioning of inhaled air and the gaseous exchange that occurs in the lungs require that the air and the vascular system of the host's tissues be in close proximity. A tough, protective epithelium composed largely of dead cells, such as the skin epidermis, is therefore not an

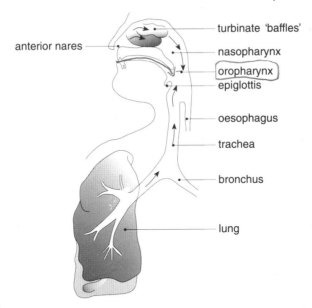

anterior nares

turbinate 'baffles'

nasopharynx

oropharynx

epiglottis

oesophagus

trachea

bronchus

lung

Figure 1.2 The respiratory tract. Arrows indicate the direction of mucociliary movement.

appropriate type of lining for most of the respiratory tract. Apart from the initial part of the tract (the anterior nares) the respiratory tract is lined with a relatively delicate epithelium which could be easily damaged by the presence of colonizing microbes. Inhaled air contains numerous microbes associated with dust and small particles of dried secretions (droplet nuclei) ejected from the nasal passages, throat and oral cavity of other humans during sneezing, coughing or loud talking. It has been estimated that air inside buildings contains about 500 microbes per cubic metre. One can calculate that the average human male, with a ventilation rate of six litres per minute at rest, would inhale about eight microbes per minute (about 10 000 per day). Despite this almost continuous inoculation, colonization of the respiratory tract is kept at a low level by host mechanisms which trap foreign objects such as microbes that are present in inhaled air and remove them from the tract.

The first 'trap' through which air must pass on entering the respiratory tract is formed by the hairs lining the nostrils. These hairs filter out large particles that are visible to the unaided eye. The nasal hairs also throw the inhaled air into turbulence, and any particles 10 μm or so in diameter will be deposited on the surface of the mucosa lining the nasal passages. This is especially true in the region of the nasal cavity where the air is forced between 'baffle plates' formed by the turbinate bones. The epithelium lining this part of the respiratory tract consists of ciliated cells and mucus-secreting cells (goblet cells). Microbes deposited on this epithelial surface are trapped in the sticky mucus covering the epithelium. The mucus, together with entrapped microbes, is carried towards the back of the throat by the beating of the cilia of the epithelial cells. The speed at which material is carried to the throat is about 0.5–1.0 cm per minute. Particles smaller than 10 μm in diameter are not trapped at this stage, but can pass into the lower regions of the respiratory tract. Small particles of 5 μm or less in diameter can reach the alveoli of the lungs. The trachea, bronchi and bronchioles also have a ciliated, mucus-producing epithelium as well as subepithelial mucus-secreting glands. Microbes trapped in the mucus on these surfaces are moved upwards out of the tract and to the back of the throat by the beating of cilia on the epithelium. This clearance mechanism in the lower respiratory tract is called the mucociliary escalator. Thus microbes entrapped in mucus in either the upper or lower respiratory tract end up in the throat and are then swallowed so as to be effectively removed from the system. The areas of the throat where microbes and other foreign substances are deposited are quite rich in lymphoid tissues (e.g., the tonsils) which help the body to resist invasion by microbes. The terminal regions of the lower respiratory tract are the alveoli which do not have mucus or cilia. Their surfaces are protected from colonization, however, by mobile host cells called macrophages. These host cells have the capacity to engulf foreign material and to destroy it by intracellular digestive processes. Macrophages can move up the mucociliary escalator to the throat; this constitutes a method of removing materials which cannot be digested by macrophages from the lungs. In health, the lungs are sterile. The upper respiratory tract (anterior nares, naso- and oropharynx), however, always harbours a microflora.

The epithelium lining the anterior nares is essentially an infolding of the epidermis. Microbes inhabiting this site are thus of the same type that are encountered on the skin (Table 1.4). *Staphylococcus epidermidis* and corynebacteria are the most commonly encountered microbes. Another staphylococcal species, *S. aureus*, is also frequently detected, being present in the anterior nares of between 20–85% of individuals in various surveys. The incidence of *S. aureus* is highest in newborn infants and lowest in the aged but the reasons for this variation are not known. The incidence of *S. aureus* in the nasal passages of hospital patients is sometimes higher than that observed for nonhospitalized people. This may be due to the state of health of hospital patients, or could be a reflection of the degree of contamination of the environment with *S. aureus* in some hospitals. Knowledge of the occurrence of *S. aureus* in the anterior nares is of medical importance because some strains of this species can cause disease (Table 1.5).

The species *S. aureus* can be subdivided into smaller groupings according to susceptibility to a collection of bacterial viruses (bacteriophages). Bacteriophage strains comprising an international set of about 20 different

Table 1.4 Bacteria commonly detected in the upper respiratory tract

Bacteria from the anterior nares
 Staphylococcus epidermidis
 S. aureus
 Corynebacterium species

Bacteria isolated from the nasopharynx
 As for the anterior nares plus:
 Moraxella catarrhalis
 Haemophilus influenzae
 Neisseria meningitidis
 N. mucosa
 N. sicca
 N. subflava

Bacteria isolated from the oropharynx
 As for the nasopharynx plus:
 Streptococcus anginosus
 S. constellatus
 S. intermedius
 S. sanguis
 S. oralis
 S. mitis
 S. acidominimus
 S. morbillorum
 S. salivarius
 S. uberis
 S. gordonii
 S. mutans
 S. cricetus
 S. rattus
 S. sobrinus
 S. crista
 S. pneumoniae
 S. pyogenes (<10% of population)
 Haemophilus parainfluenzae
 Mycoplasma salivarius
 M. orale

viruses are dropped on to a plate of culture medium seeded with the staphylococcal strain to be tested. The pattern of lysis (susceptibility) of the staphylococcus is observed after an overnight incubation. Most strains of *S. aureus* are infected, and hence lysed, by more than one bacteriophage in the typing set. The susceptibility pattern of the staphylococcal strain is known as its 'phage type' and the overall procedure is called 'phage typing'. Staphylococcal infections are characterized by suppuration, dead tissues (necrosis) and a tendency for the infected area to become walled off with the formation of a pus-filled local abscess. Certain phage types are more

Table 1.5 Diseases caused by *Staphylococcus aureus*

Furuncle (boil)	Superficial skin infection that develops in a hair follicle, sebaceous gland or sweat gland
Stye	Infection at the base of an eyelash
Carbuncle	Abscess developing after infection spreads from a furuncle to subcutaneous tissues
Bullous impetigo	Superficial skin infection characterized by large blisters. Caused especially by phage type 71
Paronychia	Infection of the soft tissue around the nails
Deep lesions	Osteomyelitis, arthritis, deep-seated abscesses due to blood-borne spread of *S. aureus* from skin lesions
Wound infections	Umbilical stump of newborn infants, surgical wound infections
Scalded skin syndrome	Blistering and desquamation of sheets of skin due to toxin produced by some strains of *S. aureus*
Toxic shock syndrome	Associated with the use of certain highly absorbent intravaginal tampons. High fever, vomiting, diarrhoea, sore throat and muscle pain, sometimes progressing to severe shock with evidence of kidney and liver damage
Food poisoning	Vomiting and diarrhoea following ingestion of food in which toxin-producing staphylococci have multiplied

likely to produce infection than others. In many infections, special circumstances (predisposing conditions) must occur before the bacteria can successfully establish in tissues and produce disease.

Subdivision of *S. aureus* into smaller groups is useful in helping to trace the spread of this organism in a hospital environment, or in determining the persistence of a particular strain in the nasal cavity. It is apparent from such studies that individuals who harbour *S. aureus* belong to one of three categories:

1. 'persistent carriers' – those who harbour a specific strain for prolonged periods;
2. 'occasional carriers' – those who sporadically harbour *S. aureus*;
3. 'intermittent (transient) carriers' – those who harbour one staphylococcal strain for a certain period and then harbour a different strain.

Colonization of the nasal passage by one *S. aureus* strain can prevent establishment of a different strain in that site. This observation was put to practical use in 1961 by Henry Shinefield and colleagues during an outbreak of infection due to *S. aureus* phage type 80/81 in a hospital nursery. They found that a nurse tending the infants in the nursery was a carrier of the 80/81 strain. The investigators observed that only infants under 24 hours of age became colonized by the 80/81 strain. Older infants apparently were not susceptible because they had acquired a different phage type of

S. aureus prior to being handled by the nurse. Shinefield and colleagues intentionally colonized the umbilical stump and nasal mucosa of newborn infants with a strain of *S. aureus* (502A) isolated from another nurse. Strain 502A prevented colonization of the babies by strain 80/81 and eliminated the disease-producing strain from the nursery. The mechanism by which strain 502A prevented the establishment of 80/81 is not known.

The nasopharynx harbours the same types of Gram-positive bacteria that are found in the anterior nares, but Gram-negative bacteria are also present (Table 1.4). *Moraxella catarrhalis* is a Gram-negative coccobacillus which is sometimes involved in the production of middle ear infections in children as well as in chronic bronchitis.

The genus *Haemophilus* is a group of Gram-negative bacilli whose members are aerobic or facultatively anaerobic. The species *Haemophilus influenzae* contains both encapsulated (polysaccharide) and nonencapsulated strains. Most of the strains harboured in the healthy respiratory tract are nonencapsulated. Encapsulated strains are sometimes present in the nasopharynx, however, and they can be subdivided into six serotypes (a to f) according to the antigenicity of their capsular polysaccharide. One of the serotypes (type b) is responsible for the causation of some serious infections of the respiratory tract and associated tissues (Table 1.6). Non-encapsulated strains are sometimes involved in the production of otitis media, acute and chronic sinusitis, and in the exacerbation of chronic bronchitis.

Gram-negative diplococci belonging to the genus *Neisseria* can be isolated from the posterior nasopharynx at the junction of the hard and soft palates. Four species of neisseria may be present (*N. meningitidis*, *N. sicca*, *N. subflava*, *N. mucosa*), but the majority of studies have centred on *Neisseria meningitidis* (the meningococcus) which is a causative agent of meningitis. The meningococcus can be detected in about 5% of healthy humans, the frequency of detection increasing to at least 20% during epidemics. It appears that the meningococcus is a member of the normal microflora of some individuals. The bacteria only rarely invade tissues and spread to other body sites, except when certain predisposing conditions (e.g. a preceding or concurrent viral infection of the upper respiratory tract) prevail.

The oropharynx contains an extensive collection of bacteria (Table 1.4), many of which are also detectable in the anterior nares and the nasopharynx. The numerically dominant organisms in the oropharynx, however, are Gram-positive cocci belonging to the genus *Streptococcus*. The generic name reflects the tendency of these bacteria to form chains of cocci. The streptococci can be placed in broad categories according to the presence or absence of enzymes (haemolysins) which act on red blood cells. The majority of streptococci isolated from the healthy oropharynx belongs to the alpha haemolytic or nonhaemolytic categories. 'Alpha haemolysis'

Table 1.6 Infections caused by *Haemophilus influenzae* (type b)

Acute epiglottitis	Sudden onset with fever, sore throat, hoarseness, barking cough. The epiglottis is swollen, inflamed, cherry-red, protruding into the airway. This infection is treated as a medical emergency with emphasis on the maintenance of an airway.
Meningitis	Often preceded by signs and symptoms of upper respiratory tract infection. Five to 10% mortality, even with adequate therapy, and many survivors have neurological damage.
Cellulitis	A tender swelling of the cheek or the areas around the eyes.
Arthritis	A major cause of purulent arthritis in children, particularly in those less than two years of age.

refers to a greenish zone surrounding the streptococcal colonies when cultured on blood-containing agar medium. The alpha haemolytic streptococci are sometimes referred to as the 'viridans' streptococci, a term also related to the green zone of haemolysis. Several species of viridans or non-haemolytic streptococci occur in the oropharynx (Table 1.4). These streptococci are a major cause of infections involving the cardiac endothelium (including the heart valves): 50–70% of cases of bacterial endocarditis are due to these bacteria.

Beta-haemolytic (complete destruction of red blood cells surrounding colonies on blood agar) streptococci may also be present in the oropharynx. *Streptococcus pyogenes* is of major interest from the medical point of view because this species is the causative agent of streptococcal pharyngitis and the skin infection impetigo. Serious noninfectious sequelae to streptococcal infection can also occur. Rheumatic fever can follow streptococcal pharyngitis, and acute glomerulonephritis can follow pharyngitis or impetigo. *Streptococcus pneumoniae* (the pneumococcus), a cause of pneumonia, is present in the oropharynx of 10–30% of individuals.

As outlined in the preceding paragraphs, many of the members of the normal microflora of the respiratory tract have the potential to produce disease. Certain predisposing conditions must exist, however, for pathological processes to occur due to the activities of members of the normal microflora. In general, the predisposing factors involve a decrease in the effectiveness of host defence mechanisms and/or tissue damage with loss of epithelial integrity (Table 1.7).

1.3 ORAL CAVITY

A striking feature of this ecosystem is the association of microbial cells with oral surfaces. Any microbe entering the oral cavity, but which lacks the

Table 1.7 Examples of conditions predisposing to infections involving respiratory tract bacteria

Disease	Causative agent	Predisposing factors
Pneumonia	S. aureus	Damage to lung or tracheobronchial tree due to virus infection, aspiration of gastric contents. Leucocyte defects or general reduction in host defences: diabetes, malignancy, old age, steroid or cytotoxic therapy, alcoholism.
	S. pneumoniae	Viral respiratory infection, chronic underlying disease: diabetes, alcoholism, renal disease, some malignancies. Damage to bronchial epithelium: smoking, air pollution. Respiratory dysfunction: alcoholic or narcotic intoxication, anaesthesia, trauma.
Meningitis	N. meningitidis	Viral respiratory infection. Peak incidence = six months to two years of age: period when transplacental antibody concentration has decreased but before appearance of acquired antibody.
	H. influenzae type b	Peak incidence = 6–18 months. Period of low immunity analogous to N. meningitidis. Upper respiratory tract infection: nasopharyngitis, sinusitis, otitis media (? viral).
Bronchitis	H. influenzae (nonencapsulated)	Exacerbation of chronic bronchitis initially due to smoking or other factors. Impairment of clearance mechanisms.
Bacterial (infective) endocarditis	Viridans streptococci	Transient bacteraemia with streptococci is common. Cardiac abnormalities (valvular insufficiency, stenosis, intracardiac shunts) result in turbulence of intracardiac blood flow that leads to further damage to endothelial surfaces and facilitates platelet and fibrin deposition. These factors produce potential sites for colonization and infection.

ability to associate with oral surfaces or the surfaces of other attached microbes, is quickly removed from the ecosystem by the flushing action of the flow of saliva continuously bathing the mouth. Saliva enters the oral cavity via ducts from the parotid, submandibular and sublingual glands and flows over all of the oral surfaces before being swallowed. Flow rates of saliva have been reported to be between 0.5 and 111 ml per hour, but vary over a 24-hour cycle. Even some of the microbial cells attached to the oral surfaces are dislodged by the flow of saliva, resulting in about 7×10^8 bacteria per millilitre.

A great variety of surfaces is present within the oral cavity: the smooth squamous epithelium of the cheeks, the tough epithelium of the tongue dorsum and gums (gingiva), and the smooth enamel surfaces of the teeth. Microhabitats exist within these general surfaces. Tooth surfaces, for example, vary from the smooth buccal (adjacent to the cheek) or lingual (adjacent to the tongue) surfaces which are exposed to a mechanical cleansing effect when food is chewed, to the more protected environments afforded by the approximal surfaces between the teeth, and the pits and fissures of the biting (occlusal) surfaces.

The gingival crevice or sulcus (a small space 1–3 mm deep between the gum and tooth surfaces) provides an anaerobic environment whose chemical composition is different from that of saliva-bathed areas. This is due to the presence of crevicular fluid, which resembles serum in composition.

The different oral surfaces and microhabitats select for microbes of appropriate attributes. The cheek epithelium is colonized predominantly by streptococcal species, mainly *Streptococcus mitis* and *Streptococcus sanguis*. There are between five and 25 bacteria adhering to each cheek epithelial cell.

The palate surface of 30–40% of healthy humans is colonized by an aerobic, Gram-negative organism which has gliding motility. The bacterium is in the form of filaments (2–10 μm in width and 50 or more μm in length) composed of closely apposed crescent-shaped cells. The filaments have a concave dorsal surface and a convex ventral surface. The microbes attach to the oral surface by their convex surface, perhaps aided by a capsule. These bacteria belong to the genus *Simonsiella* and may also be present on the cheek epithelium, at the gingival margin, on the floor of the oral cavity and on the tongue.

The tongue dorsum has a large surface area because it is covered with crypts and papillae. There are about 100 bacteria per tongue epithelial cell, the numerically predominant types being *Streptococcus salivarius*, *Veillonella* species (Gram-negative anaerobic cocci), *S. mitis*, *S. anginosus* and Gram-positive filamentous bacteria. Examples of oral filamentous bacteria are listed in Table 1.8. A Gram-positive coccus, *Micrococcus mucilagenous*, is also common on the tongue. Yeasts belonging to the genus *Candida* are found on the posterior dorsum of the tongue of between 40 and 60% of healthy humans. Under certain conditions the yeasts, mainly *Candida albicans*, cause oral diseases which are collectively known as candidiasis (Table 1.9).

Tooth surfaces above the gingival margin (supragingival) may harbour accumulations of bacteria, mainly Gram-positive filaments and streptococcal species (*Streptococcus sanguis*, *S. mitis*, *S. gordonii*, *S. intermedius*, *S. anginosus*, *S. constellatus*, *S. oralis*, *S. vestibularis* and mutans streptococci). These streptococcal species are not present in the oral cavity until

Table 1.8 Gram-positive rods and filamentous bacteria commonly detected in the oral cavity

---

Actinomyces israelii, A. viscosus, A. naeslundii
Bacteria which can form a true mycelium, although it may be transitory.
Fermentation of glucose results in the production of acetic, lactic and succinic acids. Obligate anaerobes to microaerophilic.
Eubacterium alactolyticum, E. saburreum
Produce mixed acids including acetic, butyric, caproic, lactic and succinic from glucose. Obligate anaerobes.
Lactobacillus casei, L. acidophilus, L. plantarum, L. brevis, L. fermentum and *L. salivarius*
Lactic acid is the major product of glucose fermentation. Microaerophilic.
Bifidobacterium dentium
Obligate anaerobe producing acetic and lactic acids primarily in the molar ratio of 3 : 2, and succinic acid.
Corynebacterium matruchotii
Forms a long filament growing out of a short, fat, rod-shaped cell ('whip-handle' cell). Facultative anaerobe producing acetic, lactic and propionic acids.
Propionibacterium species
Obligate anaerobes producing acetic and propionic acids.
Rothia dentocariosa
Facultative anaerobe producing acetic, lactic and succinic acids.

---

after the eruption of the teeth, reflecting their high degree of adaptation for tooth surfaces. The mutans group of streptococci is comprised of seven species, of which four (*S. mutans, S. cricetus, S. rattus* and *S. sobrinus*) are primarily of human origin.

Bacteria may also be associated with the subgingival tooth surfaces. Gram-positive filaments and streptococci are common in this site, but the gingival crevice is noteworthy for the variety of microbes (more than 350 species have been detected) including Gram-negative anaerobic types (Table 1.10) that it contains. The gingival crevice has the lowest oxidation-reduction potential (E_h) of the oral cavity, perhaps as low as $-300\,\text{mV}$. Other Gram-negative bacteria present in the oral cavity include *Eikenella corrodens*, a facultatively anaerobic rod which pits the surface of agar plates; *Haemophilus* and *Neisseria* species. Spirochaetes belonging to the genus *Treponema* (e.g. *T. denticola, T. macrodentium, T. orale, T. socranskii* and *T. vincentii*) are present in the gingival crevice. Their numbers increase markedly under conditions of gingival disease. Indeed, *T. vincentii* is a partner of *Fusobacterium nucleatum* in the causation of acute ulcerative gingivitis.

The mechanisms by which bacteria adhere to oral surfaces have been mostly studied in relation to the formation of accumulations of microbes on tooth surfaces. An adherent mass of bacterial cells on the tooth surface is known as dental plaque. There are about 10^{11} bacteria per gram of

Table 1.9 Oral candidiasis

Acute pseudomembranous candidiasis (thrush)	White patches on mucosal surfaces. A disease of young children (no immunity) and of elderly, debilitated patients.
Acute atrophic candidiasis	Suppression of other members of the normal microflora through the use of broad-spectrum antibiotics (especially tetracyclines) allows 'overgrowth' of the oral cavity by yeasts. Mucosal surfaces become thin and inflamed.
Chronic atrophic candidiasis (denture stomatitis)	Swollen, inflamed mucosa at limits of upper dentures.
Chronic hyperplastic candidiasis	White patches (leukoplakia) at the angles of the cheeks. In a small proportion of cases the lesions become cancerous.
Chronic mucocutaneous candidiasis	A rare condition of young children and elderly men. The infection, initially in the form of oral lesions, may spread to the face, nails and abdomen.
Angular cheilitis	Erosion of the angles of the mouth, particularly in skin folds. Associated with denture wearers with concurrent chronic atrophic candidiasis.

plaque. The formation of dental plaque of an appropriate bacterial content (cariogenic plaque) is a prerequisite for destruction of tooth enamel (hydroxyapatite) by acids (mostly lactic acid) which result from the metabolism of carbohydrates by plaque-associated bacteria. The formation of plaque permits bacteria to maintain themselves securely in an environment bathed by a flow of saliva. Acids produced by the metabolism of plaque bacteria cannot diffuse quickly away from the tooth surface because of the nature of the plaque structural matrix. The relatively concentrated acids thus held at the tooth surface lower the pH sufficiently for demineralization of the tooth (loss of calcium and phosphate ions) to occur. The resulting dental destruction is known as dental caries. The mechanisms by which members of the normal microflora associate with tooth surfaces and hence initiate the formation of plaque are discussed in Section 3.2.

In addition to dental caries, members of the normal microflora of the oral cavity are involved in the causation of periodontal diseases (chronic marginal gingivitis, chronic periodontitis and juvenile periodontitis). As in the case of dental caries, the formation of dental plaque is a prerequisite for the production of these diseases. Chronic marginal gingivitis is an inflammatory condition involving the gingival margin. Good oral hygiene

Table 1.10 Gram-negative bacteria commonly detected in the oral cavity

Prevotella melaninogenica, P. intermedia, P. loescheii, P. denticola	Obligate anaerobes, pleomorphic Gram-negative rods. Saccharolytic. Some strains produce black pigmented colonies when cultured on an agar medium containing blood.
Porphyromonas gingivalis, P. asaccharolytica, P. endodontalis	Obligate anaerobes, pleomorphic Gram-negative bacilli. Asaccharolytic. Colonies are pigmented black on blood agar.
Fusobacterium nucleatum, F. naviforme, F. russii, F. periodonticum, F. alocis, F. sulci	Obligate anaerobes (bacilli) producing butyric acid, without iso-butyric or iso-valeric acids, as the major fermentation product from glucose.
Leptotrichia buccalis	Obligate anaerobe (bacillus) producing lactic acid as the major fermentation product. May appear Gram-positive in young cultures.
Selenomonas sputigena, S. flueggei	Obligate anaerobes producing acetic, propionic and lactic acids. Motile by means of a tuft of flagella emerging from one side of the cell.
Capnocytophaga ochracea, C. sputagena, C. gingivalis	Grow in air or anaerobically with 10% carbon dioxide. Fusiform morphology with gliding motility.
Campylobacter rectus, C. curvus (previously known as *Wolinella*)	Obligate anaerobes, curved, helical or straight rods. Rapid, darting motility by means of single polar flagellum. Hydrogen and formate are electron donors and are used as energy sources. Fumarate and nitrate are used as electron acceptors.
Veillonella parvula, V. atypica, V. dispar	Gram-negative anaerobic cocci. Obtain energy from fermenting pyruvic and lactic acids. End products of fermentation are acetic and propionic acids, carbon dioxide and hydrogen.

which removes or reduces the extent of plaque formation in the vicinity of the gingival margins prevents, or eradicates, the disease. The amount of plaque which accumulates at the gingival margin is not the sole criterion for the production of the disease: the bacterial content of the plaque is also important. Studies of chronic marginal gingivitis have been aided by the fact that the condition can be easily produced in human volunteers by requiring the individuals to refrain from any oral hygiene procedures for a period of up to 10 days. The bacterial composition of subgingival plaque can be determined in such individuals and changes in the microbial community observed. As the plaque matures, a heterogeneous collection of microbes is found. Predominant bacteria include *Actinomyces naeslundii, A. odontolyticus, A. israelii, Veillonella parvula, Porphyromonas* and *Prevotella* species. The plaque community is dominated during the period of inflammation, however, by *Fusobacterium nucleatum, Peptostreptococcus micros, Streptococcus anginosus/constellatus*, lactobacilli, *Prevotella intermedia* and *Porphyromonas gingivalis* in addition to actinomyces and veillonella. The correlation of certain plaque bacteria with specific disease states, as is seen in the case of chronic marginal gingivitis, has given rise to the specific plaque hypothesis. Thus 'cariogenic plaque' can be characterized as consistently containing acid-producing/acid-tolerant/ extracellular polysaccharide-producing microbes; and 'periodontal plaque' as containing Gram-negative anaerobes. The subgingival plaque bacteria do not invade the gingival tissues. Their action in producing the disease state can be considered to be indirect in that the inflammation is apparently due to the penetration of extracellular microbial products into the gingival tissue. Examples of such substances include bacterial polypeptides, endotoxin, proteolytic enzymes, organic acids and toxins active against leucocytes.

Chronic periodontitis is an inflammatory disease involving not only the gingival tissues but also the alveolar bone and periodontal fibres. In chronic periodontitis, the epithelium at the bottom of the gingival crevice gradually migrates down the tooth. The deepening of the gingival crevice results in the formation of periodontal pockets, which may be up to 11 mm deep. Together with resorption of alveolar bone, this process leads to loss of the supporting tissues of the tooth, and the tooth may eventually fall out. This relatively common disease is the major cause of tooth loss after the age of 25 years. Chronic periodontitis is not due to invasion of the gingival tissue by microbes but, as seen in the case of chronic marginal gingivitis, is due to the production of extracellular products by plaque bacteria. Just which bacteria are responsible is still a matter of debate. Gram-negative anaerobic rods (*Porphyromonas gingivalis*, prevotella, *Bacteroides forsythus, Fusobacterium nucleatum*, campylobacter), spirochaetes, eikenella and actinomyces appear to be important organisms in the causation of the disease.

Juvenile periodontitis (periodontosis) is a rare disease affecting adolescents. A distinct type of alveolar bone loss occurs, often rapidly, involving the first permanent molars and the incisors. While this disease may involve hereditary factors, specific bacteria may be associated with the subgingival plaque of juvenile periodontitis. Bacterial numbers are lower in the periodontal pockets than is the case with chronic periodontitis. Two microbial types, according to some studies, seem to be particularly associated with juvenile periodontitis pockets: *Capnocytophaga ochracea* (Gram-negative carbon dioxide-requiring rods, hence 'capno-': with gliding motility) and

Actinobacillus actinomycetemcomitans (Gram-negative capnophilic rods, non-motile).

In all of these periodontal diseases, host factors must be operating in the production of the condition. The presence of plaque in the area of the gingival margin does not always result in gingival inflammation. Chronic marginal gingivitis sometimes, but not always, develops into chronic periodontitis. Juvenile periodontitis shows familial clustering. Present knowledge indicates that immunological factors influence the gingival response to microbially-produced substances. Variation in the functioning of such factors between individuals may explain differences in susceptibility to disease.

Oral microflora members are also implicated in infections of tooth pulp (root canal) and periapical tissues. Infection of living pulp by facultatively anaerobic bacteria originating from dental caries causes inflammation of the superficial pulp layers. If the pulp dies, however, the infection can spread through the root canal and result in apical periodontitis. Obligately anaerobic bacteria are involved (*Prevotella, Porphyromonas, Fusobacterium, Peptostreptococcus*) and persistent infection may result in abscess formation and osteomyelitis.

1.4 OESOPHAGUS

The oesophagus has been found to contain a normal microflora in a number of species of mammals (Table 1.11) but there is no evidence to suggest that this is so for humans. The communities present in the oesophagus of other animals have not been well studied but they are of interest because the microbes associate with the oesophageal epithelium forming clumps or layers of bacterial cells on the tissue surface.

1.5 GASTROINTESTINAL TRACT

The gastrointestinal tract is specialized to perform the functions of the digestion of food (stomach and upper small bowel), the absorption of nutrients (small bowel) and water salvage and storage of wastes (large bowel). The tract is therefore lined by specialized tissues and contains secretions, some of which originate in the biliary or pancreatic fluids, so that these functions can be performed. Only small amounts of air are swallowed and any oxygen from this source, or that diffuses from tissues lining the tract, is reduced by the metabolism of the microbial inhabitants. The lower (distal) regions of the human gastrointestinal tract are highly reducing environments with negative oxidation–reduction potentials (about $-200\,\text{mV}$ in the human colon). The chemical and physical nature of the

Table 1.11 Oesophageal microbes

Animal host	Microflora	Demonstrated by
Mouse	Lactobacilli	Culture and microscopy
Calf	Lactobacilli	Culture and microscopy
Pig	Lactobacilli	Culture and microscopy
Rat	Gram-positive cocci	Microscopy
Guinea-pig	Gram-positive cocci	Microscopy
Deer	Rod-shaped bacteria	Microscopy

gastrointestinal tract changes, therefore, as one proceeds from the stomach to the anus. These changes in environmental conditions are reflected by changes in the nature of the microflora inhabiting each site.

The human stomach and the first two-thirds of the small bowel contain only small numbers of microbes: a maximum of 10^4 per millilitre of gastric or intestinal contents. It should be noted, though, that residents of tropical countries may harbour, for unknown reasons, larger microbial populations in these regions. Microbial numbers are restricted in these areas because of the pH of the stomach contents (as low as pH 2) and the relatively swift flow (transit time of 4–6 hours) of the digesta through the stomach and small bowel (Figure 1.3). Patients who secrete abnormally low amounts of hydrochloric acid (achlorhydria) into the stomach have much larger numbers of microbes in that organ (about 10^6 per millilitre of contents). Similarly, abnormalities of the small bowel that slow the flow of the digesta

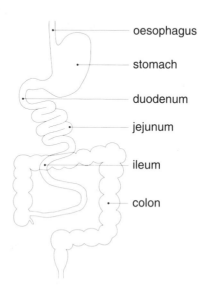

oesophagus

stomach

duodenum

jejunum

ileum

colon

Figure 1.3 The gastrointestinal tract.

result in increased colonization because the microbes are not washed out of that part of the intestine before they have time to multiply. The principal microbial types encountered in the stomach and upper small bowel are lactobacilli and streptococci. They are acid-tolerant bacteria which, unlike the majority of microbes entering the tract in food, can survive passage through the stomach. The lactobacilli and streptococci are carried into the gastrointestinal tract from the oral cavity and pharynx by saliva. Such microbes are considered to be merely passing through the upper gastrointestinal tract and are described as transients.

Until recently, the human stomach was considered, in the case of healthy subjects, never to be colonized by microbes, but the long-term association of a spiral-shaped, motile bacterium called *Helicobacter pylori* with the mucosal surface of the stomach antrum (Figure 1.4) has altered this view. Based on the presence of serum antibodies that react with helicobacter antigens, about 50% of human subjects have contact with the bacteria at some stage of their lives. The colonization of the antral surface by helicobacter does not always produce symptoms of disease, suggesting that these bacteria are normal microflora members. *Helicobacter pylori*, however, is clearly involved in the causation of inflammation of the stomach (chronic active gastritis) and in the formation of peptic ulcers (gastric or duodenal ulcers) which argues for these bacteria to be considered pathogens rather than normal microflora. Further knowledge of these microbes and their relationship with their human host is required before their exact ecological status can be determined.

The last third of the small bowel, the ileum, harbours larger numbers of microbes than are found in the upper regions of the gastrointestinal tract. The passage of the digesta (intestinal motility) is slower in the ileum and microbial colonization occurs. Lactobacilli and enterococci ('faecal streptococci') are present but Gram-negative facultatively anaerobic bacteria,

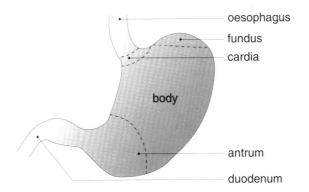

oesophagus

fundus

cardia

body

antrum

duodenum

Figure 1.4 Stomach regions.

Table 1.12 The family *Enterobacteriaceae*

Gram-negative rods, 0.3–1.0 × 1.0–6.0 μm, motile by means of peritrichous flagella or non-motile. Non-sporing. Facultative anaerobes. Nitrate usually reduced to nitrite. Usually grow on MacConkey's medium. Genera: *Cedecea, Citrobacter, Edwardsiella, Enterobacter, Erwinia, Escherichia, Ewingella, Hafnia, Klebsiella, Kluyvera, Leclercia, Leminorella, Morganella, Pantoea, Proteus, Providencia, Rahnella, Salmonella, Serratia, Shigella, Tatumella, Trabulsiella, Yersinia, Yokenella.*
Escherichia coli: Gram-negative rods, facultatively anaerobic. Citrate not utilized. Carbohydrates fermented to lactic, acetic and formic acids. Part of formic acid is split by a complex hydrogenylase system to give equal amounts of carbon dioxide and hydrogen. Lactose is fermented by most strains but fermentation can be delayed or absent. Motile by means of peritrichous flagella or nonmotile.

---

including members of the family Enterobacteriaceae, such as *Escherichia coli* (Table 1.12) and obligate anaerobes (*Bacteroides, Veillonella, Clostridium*) are common. Bacterial population levels of about 10^8 per millilitre of contents are attained.

The large bowel (colon) is the main site of microbial colonization in the human gastrointestinal tract. Intestinal motility is quite slow here (transit time of up to 60 hours) and large numbers of bacteria are present. There are about 10^{10} bacteria per gram (wet weight) of intestinal contents or faeces. Microbial cells make up about 55% of solids in colon contents. Forty species are commonly detected, but at least 400 species have been isolated in bacteriological studies of human faeces (Tables 1.13–1.15). The numerically predominant bacteria are obligate anaerobes which are 100–1000 times more numerous than facultative anaerobes such as *E. coli*.

The large bowel microflora, in theory at least, may be important in salvaging energy from dietary components that have passed undigested through the small bowel. Certainly, these dietary components are sources of nutrients for large bowel inhabitants. The large numbers of anaerobic bacteria that require a source of fermentable carbohydrate for growth mean that extensive degradation of carbohydrates occurs in the human colon. Abigail Salyers has estimated that 30–45 g of carbohydrate must be fermented each day to replace the bacterial cells lost from the ecosystem in the faeces. Complex carbohydrates that may serve as sources of fermentable substrates for intestinal microbes are listed in Table 1.16. The microbes which degrade complex carbohydrates are important members of the intestinal ecosystem because by degrading polysaccharides they release simpler molecules (carbohydrate residues and fermentation products such as short chain fatty acids) which other types of microbe can utilize.

Between 10 and 30% of protein entering the small bowel is not absorbed but passes into the colon. Microbial proteases degrade the proteins to

Table 1.13 The 25 most prevalent bacterial species in the faeces of human subjects consuming a Western diet (10^{9-10} bacteria per gram wet weight)[a]

1. *Bacteroides vulgatus*
2. *Bacteroides* species, other
3. *Bacteroides fragilis*
4. *Bacteroides thetaiotaomicron*
5. *Peptostreptococcus micros*
6. *Bacillus* species (all)
7. *Bifidobacterium adolescentis* D
8. *Eubacterium aerofaciens*
9. *Bifidobacterium infantis*, other
10. *Ruminococcus albus*
11. *Bacteroides distasonis*
12. *Peptostreptococcus intermedius*
13. *Peptostreptococcus* sp. 2
14. *Peptostreptococcus productus*
15. *Eubacterium lentum*
16. Facultative streptococci, other
17. *Fusobacterium russii*
18. *Bifidobacterium adolescentis* A
19. *Bifidobacterium adolescentis* C
20. *Bacteroides clostridiiformis* ss. *clostridiiformis*
21. *Peptostreptococcus prevotii*
22. *Bifidobacterium infantis* ss. *liberorum*
23. *Clostridium indolis*
24. *Enterococcus faecium*
25. *Bifidobactrium longum* ss. *longum*

[a](Source: Finegold *et al.* (1974) Effect of diet on human fecal flora: comparison of Japanese and American diets. *American Journal of Clinical Nutrition* **27**, 1456–69.)

peptides which can be further broken down by the majority of intestinal bacteria by virtue of the production of peptidases. Amino acids may be utilized for protein synthesis by microbial cells or degraded by deamination or decarboxylation. Many intestinal bacteria can utilize ammonia as a nitrogen source.

Ammonia in the intestinal tract results from the breakdown of amino acids, but particularly from the hydrolysis of urea. Urea diffuses from the blood into the intestinal lumen where it is hydrolysed by bacterial ureases. As much as 14% of the strains comprising the intestinal community are able to degrade urea, resulting in the hydrolysis in the large bowel of about 25% of the total daily urea excretion of humans. It is estimated that this amount of urea hydrolysis would represent a daily production of 3–4 g of ammonia. The major urease producer in the human large bowel is an obligately anaerobic Gram-positive coccus, *Peptostreptococcus productus*.

While dietary components are important nutritionally for large bowel microbes, molecules originating in the host's tissues are also a source of nutrients. These molecules include proteins present in digestive secretions,

Table 1.14 Characteristics of bacterial genera commonly detected in human faeces

Bacteroides Gram-negative, nonspore-forming bacilli. Obligate anaerobe. Metabolic products include combinations of acetic, succinic, lactic, formic or propionic acids. If *N*-butyric acid is produced, isobutyric and isovaleric acids are also present.
Bifidobacterium Gram-positive, nonspore-forming, nonmotile bacilli sometimes with club-shaped or spatulated extremities. Obligate anaerobe. Acetic and lactic acids are produced primarily in the molar ratio of 3 : 2. Glucose is degraded exclusively and characteristically by the fructose-6-phosphate 'shunt' metabolic pathway.
Clostridium Gram-positive bacilli that form endospores. Obligate anaerobe.
Enterococcus Gram-positive cocci. Facultative anaerobe. Lancefield group D. Can grow in 6.5% NaCl broth and in broth at pH 9.6.
Eubacterium Gram-positive bacilli, nonspore-forming. Obligate anaerobe. Produces mixtures of organic acids including butyric, acetic and formic acids.
Fusobacterium Gram-negative, nonspore-forming bacilli. Obligate anaerobe. *N*-butyric acid is produced, but isobutyric and isovaleric acids are not.
Peptostreptococcus Gram-positive cocci. Obligate anaerobe. Can metabolize peptone and amino acids.
Ruminococcus Gram-positive cocci. Obligate anaerobe. Amino acids and peptides are not fermented. Fermentation of carbohydrates produces acetic, succinic and lactic acids, ethanol, carbon dioxide and hydrogen.

Table 1.15 Some other bacterial species detected in human faeces

Acidaminococcus fermentans	10^{7-8a}
Bacteroides ovatus *B. uniformis* *B. coagulans* *B. eggerthii* *B. merdae* *B. stercoris*	10^{9-10}
Bifidobacterium bifidum *B. breve*	10^{8-9}
Clostridium cadaveris *C. clostridioforme* *C. innocuum* *C. paraputrificum* *C. perfringens* *C. ramosum* *C. tertium*	10^{8-9}
Coprococcus cutactus	10^{7-8}
Enterobacter aerogenes	10^{5-6}
Enterococcus faecalis	10^{5-6}
Escherichia coli	10^{6-7}

Table 1.15 *Continued*

Eubacterium limosum *E. tenue*	10^{8-9}
Fusobacterium mortiferum *F. naviforme* *F. necrogenes* *F. nucleatum* *F. prausnitzii* *F. varium*	10^{6-7}
Klebsiella pneumoniae *K. oxytoca*	10^{5-6}
Lactobacillus acidophilus *L. brevis* *L. casei* *L. fermentum* *L. leichmannii* *L. minutus* *L. plantarum* *L. rogosae* *L. ruminis* *L. salivarius*	10^{7-8}
Megamonas hypermegas	10^{7-8}
Megasphaera elsdenii	10^{7-8}
Methanobrevibacter smithii *Methanosphaera stadtmaniae*	undetectable – 10^9
Peptostreptococcus asaccharolyticus *P. magnus* *P. productus*	10^{8-9}
Proteus mirabilis *P. morgannii*	10^{5-6}
Veillonella parvula	10^{5-6}

[a]Approximate number of viable cells per gram (wet weight) of faeces, but a 10^4 variation can occur between subjects.

carbohydrates and proteins in mucus, and the components of epithelial cells which are continuously lost (extruded) from the epithelium into the lumen of the gastrointestinal tract. Information on the contribution of host secretions to potential nutrients for microbes is provided in Table 1.17.

In general, the microbial degradation of complex molecules in the large bowel can be thought of as a sequential process in which a number of microbial types play their part. The cleavage of polymers into smaller subunits provides substrates which other microbes can utilize in their metabolism. The removal of certain chemical groups or molecules from complex structures by some microbial types may make other substrates

Table 1.16 Complex carbohydrates in the human colon

Source	Type of carbohydrate
Plant structural material	Cellulose
	Laminarin
	Xylan
	Arabinogalactan
	Pectin
Food additives	Guar gum
	Gum arabic
	Alginate
Vegetables, flour	Amylose
	Amylopectin
Host cells	Mucopolysaccharides
Mucus	Mucins

Table 1.17 Host secretions as sources of nutrients for intestinal microbes

Daily output of secretions in the human digestive tract (l)

Saliva	1.0–2.0
Gastric	0.3–1.4
Pancreatic	0.2–0.8
Bile	0.3–1.1
Colonic	0.02–0.2

Daily secretion of minerals into the human digestive tract (g)

	Sodium	Potassium	Chloride	Bicarbonate
Saliva	0.21	0.12	0.10	0.54
Gastric	4.68	0.86	5.54	4.11
Bile	2.90	0.02	1.35	1.06
Pancreas	1.61	0.008	1.75	0.61

Daily secretion of protein into the small bowel (g)

Intestinal secretions	10–30
Desquamated cells	20–30
Plasma protein	1–2

(Source: Hove, E.L. (ed.) (1974) *Proceedings of Symposium on Absorption from the Alimentary Tract*, Nutrition Society of New Zealand.)

available for other bacteria. The degradation of complex substrates in the intestine is therefore a cooperative process which involves crossfeeding between various microbial types (microbial consortia). The significance of these processes to the host is discussed in Section 4.1.

1.6 UROGENITAL TRACT

Most of the urinary tract of humans, in health, is free of microbial life. Only the distal urethra (Figures 1.5 and 1.6) harbours a microflora composed of

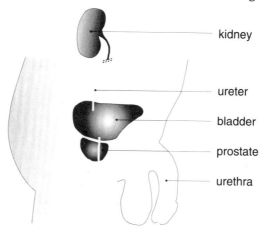

kidney

ureter

bladder

prostate

urethra

Figure 1.5 The male urinary tract.

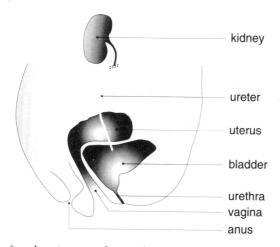

kidney

ureter

uterus

bladder

urethra
vagina
anus

Figure 1.6 The female urinary and genital tracts.

bacteria that have entered the tract due to contamination of the tissue surfaces surrounding the urethral orifice. The microbes have their origin in the faeces, the vaginal secretions and on the skin. Urethral organisms commonly include *Escherichia coli*, lactobacilli and staphylococci. The microbes do not succeed in colonizing very far up the urethra, however, mostly due to the flushing action of urine flowing through the tract. In health, voided urine contains bacteria washed from the urethral epithelium, the numbers of *E. coli* never exceeding 10^5 per millilitre. Other factors that prevent urinary tract colonization by microbes in some individuals include urine pH, high osmolality and concentration of urea. Antibody production,

Table 1.18 Predisposing conditions to urinary tract infection

Condition	Explanation
Congenital abnormalities, calculi (stones), enlarged prostate	Strictures or partial blockages which impede the free flow of urine through the urinary tract
Pregnancy	Diminished bladder wall tone: bladder does not empty completely leaving 'sump' of urine which can contain a reservoir of bacteria Decreased ureteral peristalsis; dilatation of ureters and renal pelvis; reflux of urine which can aid microbial passage up the tract
Catheterization	Introduction of microbes into urinary bladder when the catheter is inserted
Indwelling catheter	Provides foreign structure which aids microbial colonization of urethra and migration of microbial cells to bladder

and an ill-defined antibacterial property of the urinary bladder wall, may also protect against infection.

Infections of the urinary tract can occur when certain predisposing conditions exist (Table 1.18). Most of these infections are caused by *E. coli* of faecal origin, but *Staphylococcus saprophyticus* (usually inhabiting the skin of the feet) is common in infections of young women. Urinary tract infections are about 14 times more common in females than in males. This is partly due to the difference in length of the urethra between males (20 cm) and females (5 cm) which makes the bladder of women more easily accessible to microbes. Changes in the functioning of the tract under the influence of pregnancy, and massaging of the urethra during sexual intercourse, are additional factors contributing to the higher prevalence of urinary tract infection in women than in men.

The mucosa of the external genitalia of both males and females harbours *Mycobacterium smegmatis* (a Gram-positive bacillus with a high content of waxes and lipids in its cell wall). The male genital tract, apart from the last third of the urethra, is bacteriologically sterile. Microbial colonization is minimized by the presence of antibacterial substances (spermine, zinc) in prostatic fluid, in addition to the protective mechanisms outlined in relation to the urinary tract. Part of the female genital tract (the vagina) is, however, always colonized by microbes. Normal vaginal secretions contain about 10^8 bacteria per millilitre. The normal microflora of the vagina is the least well understood collection of microbes inhabiting the human body. This is partly due to the variability in the quality of sampling and

Table 1.19 Vaginal lactobacilli

Lactobacillus acidophilus
L. fermentum
L. casei subspecies *rhamnosus*
L. cellobiosus
L. plantarum
L. brevis
L. delbrueckii
L. salivarius
L. jensenii
L. vaginalis
L. gasseri
L. crispatus

microbiological procedures used in studies to date. The composition of the group of subjects examined in such studies, however, is probably the main source of variation in attempting to define the normal vaginal microbes. Some studies have relied on samples collected from young women attending university health clinics, others on samples collected from patients attending hospital 'sexually transmitted disease' clinics, and yet others from women attending family planning clinics. Lifestyles must differ greatly between such groups and may explain the diversity of results appearing in the medical and scientific literature. All studies agree that lactobacilli (Table 1.19) are the numerically dominant bacteria present in the vagina of healthy women in the child-bearing age group. These microbes associate with the surface of the vaginal epithelium and their Gram-positive cells can be seen adhering to desquamated epithelial cells in microscope smears prepared from vaginal washings. Lactobacilli are common in the vagina between puberty and the menopause when the vaginal environment has an acid pH. The vaginal epithelium during these years, due to the influence of oestrogens, contains glycogen. The glycogen is utilized by either the vaginal epithelial cells themselves, or by bacterial cells. The fermentation of the glucose residues comprising the glycogen molecule leads to the accumulation of lactic acid in the vaginal environment. The acid environment of the healthy vagina (about pH 4.7) selects for acid-tolerant bacteria (the lactobacilli) which become the numerically dominant microbes. Before puberty, and after the menopause, glycogen is not present in the vaginal epithelium. The vaginal secretion is alkaline under these conditions and staphylococci and streptococci, rather than lactobacilli, are the numerically dominant bacteria. The vaginal ecosystem provides, therefore, a good example of the influence of host physiology on the composition of the normal microflora.

The results of studies of the vaginal ecosystem vary in the extent to which obligate anaerobes form part of the normal microflora of the vagina. Two

Table 1.20 The normal microflora of vaginal washings.

Microbial group	Mean number of microbes per ml of washings (log_{10})	Subjects harbouring microbes (%)
Lactobacillus	7.2	87
Staphylococcus	4.0	62
Ureaplasma	5.0	54
Corynebacterium	4.2	37
Streptococcus	4.9	25
Anaerobic Gram-positive cocci	5.9	19
Gardnerella	7.7	17
Bacteroides	5.8	14
Mycoplasma	3.7	14
Candida	4.2	14
Eubacterium	6.9	8
Anaerobic streptococcus	8.4	6
Bifidobacterium	6.3	6
Propionibacterium	5.2	4
Escherichia	4.5	4
Sarcina	6.0	2
Klebsiella	5.8	2
Fusobacterium	5.0	2

Single samples of vaginal washings were obtained from 52 women attending a family planning clinic. (Source: Cook, R., Tannock, G. and Meech, R. (1984) The normal microflora of the vagina. *Proceedings of the University of Otago Medical School* **62**, 72–4.)

patterns emerge: obligate anaerobes are not common inhabitants of the vaginal ecosystem; and obligate anaerobes are common in the vagina (Tables 1.20, 1.21). This is an important question which needs to be resolved because anaerobic bacteria appear to be one of the causes of vaginitis (diffuse or patchy inflammation of the vaginal mucosa accompanied by abnormal vaginal secretions). Cases of vaginitis can be grouped according to the nature of the causative agent of the disease: vaginal thrush caused by *Candida albicans*; vaginitis due to a protozoan *Trichomonas vaginalis*; and vaginitis to which a single causative agent cannot be ascribed (nonspecific vaginitis, also known as bacterial vaginosis). Candida vaginitis is often accompanied by irritation of the vaginal epithelium, vulval itch and a thick curd-like discharge; trichomoniasis by a purulent, malodorous discharge and vulvar itching; and nonspecific vaginitis by increased amounts of vaginal secretions which are often malodorous.

Nonspecific vaginitis has proved to be a difficult microbiological problem to solve. Earlier work linked the causation of the disease to a bacterium called, at that time, *Haemophilus vaginalis*. Later studies showed that, unlike members of the genus *Haemophilus*, *H. vaginalis* did not require X or V factors or any other type of growth factor. The organism has also been placed in the genus *Corynebacterium* because of the demonstration of

Table 1.21 The normal microflora of vaginal secretions

Microbial group	Mean number of microbes per gram of secretions (log_{10})	Subjects harbouring microbes (%)
Staphylococci	7.5	41
Streptococci	6.8	59
Lactobacilli (aerotolerant)	8.7	50
Corynebacterium	7.2	31
Escherichia	6.4	9
Anaerobic Gram-positive cocci (peptococci)	8.7	64
Anaerobic Gram-positive cocci (peptostreptococcus)	8.3	23
Veillonella	7.6	9
Anaerobic lactobacilli	8.2	45
Eubacterium	8.4	36
Propionibacterium	8.6	14
Bifidobacterium	8.6	5
Anaerobic Gram-negative bacilli	7.7	41

Single swab specimens of vaginal secretions from 17 college students and 35 specimens from 5 other subjects. (Source: Barlett, J.G., Onderdonk, A.B., Drude, E., Goldstein, C., Anderka, S.A. and McCormack, W.M. (1977) Quantitative bacteriology of the vaginal flora. *Journal of Infectious Diseases* **136**, 271-7.)

intracellular 'granules' and 'club-shaped' cells. Genetic studies of the bacteria have shown, however, that the microbe does not resemble previously described genera, and is now called *Gardnerella vaginalis*: a Gram-negative to Gram-variable small bacillus, microaerophilic, producing acetic acid as a major fermentation product from glucose. *Gardnerella* adhere to vaginal epithelial cells which, when observed in microscopically examined smears of vaginal discharge, are referred to as 'clue' cells since they are characteristic of nonspecific vaginitis.

The history of *Gardnerella* as a possible causative agent of nonspecific vaginitis is almost as varied as that of its taxonomy. Once considered to be the cause of 90% of cases of nonspecific vaginitis, then relegated to the role of co-causative agent in conjunction with anaerobes, *Gardnerella vaginalis* may, in the opinion of some, simply be a member of the normal microflora that increases in numbers when abnormal physiological conditions exist in the vagina. The same may be said of anaerobic Gram-positive spiral-shaped bacteria, placed in the genus *Mobiluncus*, which are more readily isolated from the vagina of subjects suffering from vaginitis than from healthy women.

The role of obligate anaerobes in the causation of nonspecific vaginitis has also received attention. The involvement of these bacteria is suggested by the fact that the majority of cases of nonspecific vaginitis is cleared by the use of an antibiotic, metronidazole, which is active against anaerobic bacteria but much less so against facultatively anaerobic organisms. Further evidence of the involvement of anaerobes in nonspecific vaginitis is provided by the work of Spiegel and colleagues. They found that lactic acid was the major acid present in normal vaginal fluid. Lactic acid-producing bacteria (lactobacilli and streptococci) were the predominant bacteria in these specimens. Vaginal secretions from nonspecific vaginitis patients, in contrast, contained low concentrations of lactic acid but high amounts of succinic, acetic, butyric and propionic acids. Anaerobes that produce these acids (*Prevotella bivia*, *Prevotella disiens*, anaerobic cocci) were the predominant bacteria. Following treatment with metronidazole, the signs and symptoms of nonspecific vaginitis cleared, butyric and propionic acids disappeared from vaginal samples, and lactic acid became the principal acid detected. In addition to short chain fatty acids derived from the metabolism of anaerobic bacteria, vaginal samples from nonspecific vaginitis patients contain amines. Putrescine and cadaverine, in particular, are present in abnormal secretions but not in normal vaginal samples. It has been proposed that amine production by anaerobes inhabiting the vagina contributes to the higher-than-normal pH (5.0–5.5) of vaginal secretions in nonspecific vaginitis patients and to the 'fishy' odour of such samples. Some anaerobic bacteria can degrade amino acids by decarboxylation or deamination. Bacterial decarboxylases are induced in the presence of an appropriate substrate at a pH less than 6.0. Under laboratory conditions, the enzymes are produced late in the logarithmic stage of bacterial growth when acids produced by cell metabolism accumulate in relatively high concentrations. The induction of decarboxylases is thus considered to be a means of regulating environmental pH by the production of highly basic amines from amino acids.

From a somewhat murky picture, the following sequence of events in the causation of nonspecific vaginitis can be proposed.

1. Due to unexplained physiological influences, the vaginal ecosystem changes so that lactic acid-producing and acid-tolerant bacteria are no longer the predominant members of the vaginal microflora.
2. Obligate anaerobic bacteria such as prevotella and anaerobic cocci which are already minor members of the microflora, or which are introduced into the vagina from the faeces, attain bacteriologically detectable population levels.
3. Under the acid conditions of the vagina, decarboxylases are induced in anaerobic bacteria and amines are liberated through the degradation of amino acids. Vaginal pH is raised and vaginal secretions are malodorous.
4. Amines in the vaginal ecosystem (e.g. histamine) irritate the vaginal mucosa leading to increased amounts of vaginal secretion.

The scientific understanding, or lack of clear understanding, of the causation of nonspecific vaginitis provides a good example of how basic knowledge of the normal microflora of a body site is necessary before pathological conditions can be investigated satisfactorily. The normal microflora of the vagina has not yet been adequately defined, largely because of the divergence of results obtained through variation in the nature of culture techniques, sampling methods, choice of human subject and analysis of results that have occurred throughout the course of its investigation.

FURTHER READING

Drasar, B.S. and Barrow, P.A. (1985) *Intestinal Microbiology*, American Society for Microbiology, Washington, D.C.

Drasar, B.S. and Hill, M.J. (1974) *Human Intestinal Flora*, Academic Press, London.

Hentges, D.J. (ed.) (1983) *Human Intestinal Microflora in Health and Disease*, Academic Press, New York.

Lee, A., Fox, J. and Hazell, S. (1993) Pathogenicity of *Helicobacter pylori*: a perspective. *Infection and Immunity* **61**, 1601–10.

Marples, M. (1965) *The Ecology of the Human Skin*, Charles C. Thomas, Springfield.

Marples, M. (1969) Life on the human skin. *Scientific American* **220**, 108–15.

Marsh, P. and Martin, M. (1984) *Oral Microbiology*, 2nd edn, American Society for Microbiology, Washington, D.C.

Noble, W.C. (1990) Factors controlling the microflora of the skin, in *Human Microbial Ecology* (eds M.J. Hill and P.D. Marsh), CRC Press, Boca Raton, FL, pp. 131–53.

Noble, W.C. and Somerville, D.A. (1974) *Microbiology of Human Skin*, W.B. Saunders, London.

Salyers, A.A. (1982) Enzymes involved in degradation of unabsorbed polysaccharides by bacteria of the large bowel, in *Fibre in Human and Animal Nutrition* (eds G. Wallace and L. Bell), The Royal Society of New Zealand, Wellington, pp. 135–8.

Savage, D.C. (1977) Microbial ecology of the gastrointestinal tract. *Annual Reviews of Microbiology* **31**, 107–33.

Sherris, J.C. (ed.) (1990) *Medical Microbiology*, 2nd edn, Elsevier, New York.

Shinefield, H.R., Ribble, J.C., Boris, M. and Eichenwald, H.F. (1963) Bacterial interference: its effect on nursery-acquired infection with *Staphylococcus aureus*. I. Preliminary observations on artificial colonization of newborns. *American Journal of Diseases of Children* **1051**, 646–54.

Slots, J. and Taubman, M.A. (1992) *Contemporary Oral Microbiology and Immunology*. Mosby Year Book, St Louis.

Spiegel, C.A. (1991) Bacterial vaginosis. *Clinical Microbiology Reviews*, **4**, 485–502.

2 Happy Birthday: the acquisition of the normal microflora

If microbes were capable of emotion, they would celebrate each time an infant was born. At birth, a new potential host emerges from the protected environment of the uterus, providing pristine surfaces and body cavities as sites for microbial colonization. Members of the vaginal microflora have the opportunity to contaminate the infant during its passage through the birth canal. At the moment of birth or shortly thereafter, microbes in faeces that have been expelled involuntarily by the mother during labour, together with environmental microbes, have the opportunity to colonize the neonate. Suckling, kissing and caressing of the infant after birth provide additional insurance that members of the microflora are transmitted from one generation to the next.

Microbial colonization of the neonate occurs within the 24 hours following birth. Proliferation of microbial types in body sites appears to be initially unchecked, resulting in a heterogeneous collection of microbes. Soon, however, regulatory mechanisms generated within habitats (autogenic factors) and by external forces (allogenic factors) permit the continuing presence of some microbial types in the body's ecosystems but the elimination of others. These qualitative and quantitative changes that occur in microbial populations inhabiting the young animal provide an example of a biological succession. Eventually, usually after weaning, the microbial composition of the microflora becomes more stable and the adult microflora (in ecological terms, the climax community) is attained. A small amount of evidence is available that suggests that the normal microflora, at least of humans, is influenced by genetic determinants of the host. The oral and nasal microfloras, and perhaps that of the faeces, are more similar in comparisons of twins than in the case of singleton children. Although a climax faecal microflora characteristic of humans is recognizable, variation among individuals is greater than variation between samples collected from a single individual.

2.1 ACQUISITION OF THE NORMAL MICROFLORA OF ANIMALS OTHER THAN HUMANS

Information relating to the acquisition of the normal microflora has been most easily obtained from studies of experimental animals. Access to these animals at different ages, and collection of adequate samples for microbiological analysis, are simpler than in the case of humans. This is particularly true of studies of the intestinal microflora where, in the case of humans, observations have been largely restricted to the examination of faecal samples. The sequence of events occurring during the acquisition of the gastrointestinal tract microflora of rodents has been well described.

The initial microbial inhabitants of the gastrointestinal tract of neonatal mice are of environmental (e.g. flavobacteria), or maternal (staphylococci, enterococci, *Escherichia coli*, lactobacilli) origin. Microbes originating in the general environment do not persist in the gastrointestinal tract for long, however, and by about 10 days after birth, lactobacilli, enterococci and *E. coli* are especially numerous (about 10^8 per gram of large bowel contents). Lactobacilli predominate in the proximal region of the mouse gastrointestinal tract because at least some strains can adhere to and colonize the surface of the stratified squamous epithelium lining the oesophagus and forestomach. Colonization of these surfaces results in a layer of *Lactobacillus* cells associated with the epithelium. This phenomenon is discussed further in Section 3.3. Lactobacilli shed from this layer inoculate the digesta as it flows past so that these bacteria are present throughout the length of the gastrointestinal tract. The *Lactobacillus* population of the murine gastrointestinal tract is retained at a high level throughout the remainder of the animal's life. Enterococci and *E. coli* are confined to the distal (away from the origin) regions of the intestinal tract, probably because they cannot adhere to the epithelium of the proximal (towards the origin) tract and are flushed through the small intestine at a rate that does not permit the accumulation of appreciable numbers of bacteria. The enterococci and *E. coli* form microcolonies in the mucus covering the epithelium of the distal intestinal tract of neonates. Microaerophilic, spiral-shaped bacteria also colonize the mucus of the distal intestinal tract and persist in adulthood. Obligately anaerobic bacteria do not establish in the gastrointestinal tract until the infant begins to supplement its milk diet by nibbling on solid food (an example of an allogenic factor influencing colonization). *Bacteroides* and fusiform-shaped bacteria (bacilli with tapered ends, e.g. clostridia and fusobacteria) colonize the large intestine at this stage, many types associating with the mucus layer covering the intestinal surface (Section 3.3). The obligate anaerobes attain populations of 10^{10} per gram (wet weight) of large intestine and become the most numerous of the microbial types in this region of the tract. The establishment of bacteroides and fusiforms results in a marked reduction in enterococcal and *E. coli* numbers due to the production of short chain fatty acids (particularly butyric acid) by the obligate anaerobes. These acids are inhibitory to the facultative anaerobes and provide an example of an autogenic factor influencing colonization. Populations of these latter bacteria decrease to about 10^4 per gram of large bowel and remain at this lower level thereafter. The microcolonies of enterococci and *E. coli* in the mucus layer are obliterated by the mass of fusiforms that forms a mucus-associated layer in the large intestine. Filamentous, segmented bacteria that attach by one end to the villous surface through the formation of an invagination of the enterocyte membrane (Section 3.3) colonize the ileum of mice after weaning, which usually occurs at the end of the third week after birth. In some colonies of mice, a yeast (*Candida pintolopesii*) establishes in the stomach after weaning and associates with the epithelial surface of the secretory mucosa. The biological succession is thus complete by the end of the fourth week after birth and a microflora characteristic of an adult mouse is present in the gastrointestinal tract. This microflora can be detected in adult mice throughout the remainder of their life as long as they are maintained free of stress.

The same general pattern of colonization to that observed with mice occurs in the other animal species that have been studied. *Escherichia coli* and enterococci are numerous initially but decrease as other microbial types become established. The adult microflora of different animal species differs in composition, however, possibly because of host dietary and physiological differences, although obligately anaerobic bacteria are always the numerically predominant microbes inhabiting the large intestine. The distribution of the microflora in the gastrointestinal tract is markedly different between animal species (Figure 2.1), mostly reflecting anatomical differences between hosts. Ruminants, for example, harbour an extensive microflora in the proximal part (rumen) of the gastrointestinal tract as well as in the large intestine, whereas the human stomach and small intestine contain only low numbers of microbes. There are numerous lactobacilli in the proximal region of the gastrointestinal tract of fowl, pigs, mice and rats because of the presence of stratifed, squamous epithelia to which the bacteria can adhere (Section 3.3). The climax community thus reflects the selection of specific microbial strains from a heterogeneous collection that initially invades the gastrointestinal lumen of the neonate. The selected strains possess attributes that enable them to thrive under the conditions existing in a particular type of gastrointestinal tract. Adaptation of microbes to life in association with the gastrointestinal tract of a particular animal species is well exemplified by lactobacilli: strains that associate with epithelia of fowl or of pigs will not colonize rodent epithelia, and conversely. Even more host-specific, filamentous segmented microbes from mice will not colonize the intestinal tract of rats and vice versa. Autogenic factors operating within habitats in the gastrointestinal tract are doubtless

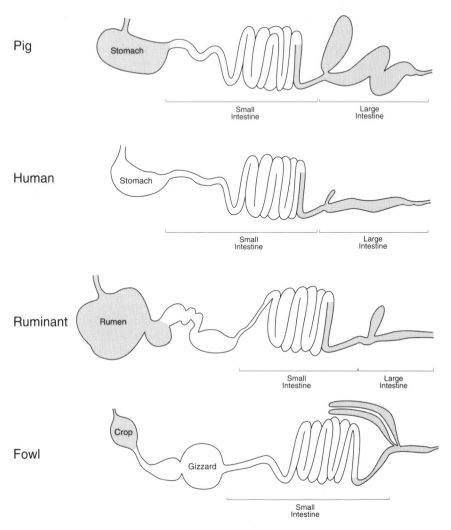

Figure 2.1 The distribution of the microflora in the gastrointestinal tract of pigs, humans, ruminants and fowl. Areas harbouring the normal microflora are indicated by shading.

of major importance in the outcome of the colonization process. However, apart from the clearly demonstrated involvement of short chain fatty acids under appropriate conditions of pH and E_h in amensalism in the large intestine, specific *in vivo* examples are unavailable. Microbial interactions in natural environments are difficult to investigate scientifically, and the results of *in vitro* experiments are notoriously unreliable indicators of microbial activities in natural habitats.

2.2 ACQUISITION OF THE NORMAL MICROFLORA OF HUMANS

2.2.1 The skin

Most attention has been directed to the acquisition of *Propionibacterium acnes* because of association of this species with acne. Propionibacteria are rarely detected on the skin prior to puberty. Their numbers increase steadily following puberty and attain what can be considered adult levels when the host is about 16 years of age. It is probable that the initiation of colonization of the skin by propionibacteria is related to the increased activity of the sebaceous glands and apocrine sweat glands and changes in the composition of sebum that commence at puberty (the increased availability of fatty acids, moisture and the altered pH of secretions).

2.2.2 The oral cavity

The neonatal oral cavity offers only epithelial surfaces for colonization so *Streptococcus salivarius*, which colonizes the tongue dorsum epithelium, is prevalent. Mutans streptococci, *Streptococcus sanguis* and other streptococci that characteristically colonize tooth surfaces are absent from the ecosystem. The oral cavity of infants, by 12 months of age, commonly harbours streptococci, staphylococci, neisseria and veillonella. Present at a lower frequency are actinomyces, lactobacilli, rothia, fusobacteria, prevotella/porphyromonas, leptotrichia, candida and corynebacteria. The occurrence of actinomyces and obligately anaerobic bacteria is greater after the eruption of the teeth, but prevotella and treponemes increase in numbers during adolescence.

2.2.3 The gastrointestinal tract

There has been controversy as to the colonization of the intestinal tract of infants. Conflicting results relating to the populations of bifidobacteria and *E. coli* in the faeces of breast milk and formula-fed infants have been published, with some studies interpreted to show that large populations of bifidobacteria only occur in breast milk-fed babies and that *E. coli* populations are lower in these infants compared to those that received a formula feed.

The 'classical' studies concerning the acquisition of the infant intestinal microflora are generally considered to be those of Tissier (1905). He divided the colonization of the intestinal tract of suckled infants into three phases. In the first phase, which consisted of the first few hours of life, the faeces were devoid of microbes. The second phase began between the 10th and 20th hour of life with the detection of a heterogeneous collection of microbial types in the faeces. After three days, by which time milk had passed through the length of the intestinal tract, the third colonization phase began. A Gram-positive bacillus became the numerically dominant microbe in the faeces at this stage as judged by microscope examination of faecal smears. The other microbial types disappeared in a fairly constant order, and from the start of the third or fourth day of life until weaning the collection of microbes in the faeces remained the same. According to Tissier, the collection seemed

> to be constituted, by microscopic examination, of only one species, *Bac. bifidus*, a strictly anaerobic bacterium. It is necessary to do a complete bacteriological examination to see that there exists, besides this species, other facultatively anaerobic bacteria in very limited numbers: *Bact. coli* (variety *commune*), *Enterocoque* and sometimes *Bact. lactis aerogenes*.
>
> *Tissier* (1905)

The faeces of infants not suckled at the breast, in contrast, continued to contain a mixture of microbial types in which no single type predominated, even after the fourth day of life. To quote Tissier,

> this usual flora is constituted like this: besides *Bac. bifidus*, *Bact. coli* (v. *commune*), *Enterocoque*, there exists in equal numbers *Bac. acidophilus*, *Bac. exilis*, more rarely *Staphylocoque blanc*, *Sarcines*, *Bact. lactis aerogenes*, *Bact. coli* (v. *communior*), *Levure blanche* and sometimes *Bac. III de Rodella* and *B. coli* (v. *typhimorphe*).
>
> *Tissier* (1905)

Tissier's turn-of-the century microscope observations of the faeces of breast milk-fed infants are still valid, but the situation regarding formula-fed babies appears to have altered with improvements to formula feeds which now resemble, but still do not exactly match, the composition of human milk. Well documented modern studies show that bifidobacteria are just as common and likely to be numerically dominant in the faeces of formula-fed as in breast-fed infants (Tables 2.1, 2.2). There is considerable infant-to-infant variation in the population size of particular bacterial genera during the first week of life in both infant groups (Table 2.3) which may have contributed to the somewhat variable interpretations of the status of the infant microflora reported in the literature. More consistent values are obtained in babies older than one week, however, and realistic comparisons between infant groups are possible. If biologically significant

Table 2.1 Incidence of selected bacterial genera in the faeces of infants aged 4–7 weeks[a]

Bacterial group	No. of infants harbouring the specified bacteria/total no. of infants examined	
	Breast milk	Formula feed
Bifidobacteria	35/35	33/35
Enterobacteria	34/35	35/35
Enterococci	33/35	35/35
Clostridia (lecithinase-producing)	5/35	18/35
Clostridia (others)	16/35	33/35

[a](Source: Benno *et al.* (1984).)

Table 2.2 Populations of selected bacterial genera in the faeces of infants aged 4–7 weeks[a]

Bacterial group	Mean bacteria per gram of faeces (log_{10})	
	Breast milk	Formula feed
Bifidobacteria	10.6	10.3
Enterobacteria	6.1	9.4
Enterococci	6.3	9.6
Clostridia	<3.0	6.4

[a](Source: Stark and Lee (1982).)

differences exist between the microfloras of breast milk or formula-fed babies, it should be possible to predict, on the basis of the microbiological study of the faeces, what the diet of the infant consisted of. Quantitative information (population sizes) about certain bacterial groups (clostridia, enterococci, *E. coli*) do appear to be indicative of the diet for infants of about four weeks of age (Table 2.2). The best prediction though, derived from data from eight published studies, is obtained with the clostridia. The clostridial population in the faeces of infants of this age is always lower in breast-fed, compared to formula-fed, infants. The significance of this difference to the infants is not known. Clostridia are possibly involved in the causation of necrotizing enterocolitis, but the peak incidence of this disease is in infants of four days of age, rather than four weeks.

It is generally believed by paediatricians that the incidence of gastro-intestinal upset is lower in breast-fed infants than in those consuming formula feeds. This difference may be due to the different physicochemical properties of human and cow's milk. The overall effect of these different properties appears to be a reduced buffering capacity of the faeces of formula-fed infants. The pH of the faeces from breast milk-fed infants tends to be lower, more than 30% of such infants having a faecal pH of less

Table 2.3 Variation in bacterial populations in infant faeces during the first week of life[a]

Bacterial group	Range of log_{10} viable counts per gram of faeces	
	Breast milk	Formula feed
Enterobacteria	<3.0−10.0	<3.0−11.0
Enterococci	3.0−10.0	<3.0−10.0
Bifidobacteria	<4.0−10.0	<4.0−10.0
Bacteroides	<3.0−10.0	<3.0−8.0

[a](Source: Stark and Lee (1982); Moreau *et al.* (1986).)

than 5. The addition of oligosaccharides (TOS: transgalactosylated oligo-saccharides) to formulas is aimed at achieving a lowering of the faecal pH since these carbohydrates pass through the small intestine without alteration but are fermented by bacteria in the large intestine. Human milk contains a large number of antibacterial, antiviral and antiparasitic factors (e.g. antibodies and complement, lactoferrin, lactoperoxidase, lysozyme, macrophages, neutrophils and lymphocytes, haemagglutinins, lipids). These factors are absent from formula feeds, or are present at low levels. Human milk also contains epidermal growth factor, nerve growth factor, somatomedin-C, insulin-like growth factor, insulin, thyroxine, cortisol, taurine, glutamine and amino sugars, all of which are believed to promote the maturation of the gastrointestinal tract in neonates. Lack of exposure of the intestinal mucosa to these molecules may delay maturation of the infant gastrointestinal tract so that it is more susceptible to upsets.

The acquisition of the gastrointestinal tract microflora is commonly described in terms of qualitative and quantitative changes at the taxonomic level of bacterial genera. Differentiation between species of a single bacterial genus, or between strains of a single species, however, reveals a more complex situation. In Japan, *Bifidobacterium breve* and *Bifidobacterium infantis* have been reported to be the numerically dominant bifidobacteria in the intestinal tract of human infants, whereas *Bifidobacterium longum* and *Bifidobacterium adolescentis* are said to be dominant in the case of adults. Enterococci inhabiting the gastrointestinal tract of mice belong to two species: *Enterococcus faecium* and *Enterococcus faecalis*. In infant mice, all of the *E. faecium* isolates ferment xylose whereas 43% of isolates from adults do not. Plasmid profiling, a technique that involves the demonstration of the plasmid molecules characteristic of a particular bacterial isolate, enables different strains of lactobacilli to be recognized. Use of this technique in examining the colonization of the neonatal piglet gastrointestinal tract by lactobacilli has permitted the observation of a complex sequence of events as piglets of a single litter matured. In this research, the pars oesophageal epithelium in the stomach was colonized by lactobacilli within 24 hours of birth. The collection of *Lactobacillus* strains, as revealed by plasmid profiling, was different in the two one-day-old piglets that were examined. At four days of age, piglets harboured different plasmid profile types of lactobacilli to those detected at 24 hours, but a similar collection of strains was present in both animals examined. Other *Lactobacillus* strains belonging to the species *Lactobacillus acidophilus* and *Lactobacillus fermentum* had colonized the pars oesophagea by seven days after birth and one particular plasmid profile type of *Lactobacillus acidophilus* had become dominant in the two animals examined. This same strain was dominant in piglets examined 14 days after birth. Even when a bacterial genus or species appears to be stably colonizing the gastrointestinal tract, therefore, changes in the composition of the strains that make up the total population can be occurring. Stability in the composition of bacterial populations may be achieved eventually in habitats, but studies involving human subjects suggest that *E. coli* populations continue to change in strain composition even in adults. While stability in species composition of the normal microflora may be common, stability at the level of bacterial strains may be less common. If this is the case, the acquisition of the normal microflora may be neverending as new strains of endogenous or exogenous origin proliferate and attain dominance in the gastrointestinal tract under the influence of particular allogenic or autogenic factors.

The biochemistry of the gastrointestinal tract is influenced by the presence of the normal microflora (see Chapter 4 for further discussion of this topic). Thus the acquisition of the normal microflora can be followed by measuring biochemical factors as well as by detecting and enumerating particular groups of microbes in the laboratory. Germfree (lacking a normal microflora) rodents have a high level of proteolytic activity in large intestinal contents compared to conventional (colonized by a microflora) animals. This is because, in the germfree animal, trypsin secreted in the pancreatic juice passes down the small intestine and remains active in the large bowel. In conventional rodents, the tryptic activity is reduced because the members of the normal microflora inactivate the proteolytic enzyme. Inactivation of tryptic activity in the caecum of rats occurs as early as 10 days after birth, indicating that the microflora appropriate for this function establishes early in the biological succession. Faeces collected from adult humans do not generally have tryptic activity. Human infants, however, usually have faecal tryptic activity which can be regularly detected until the children are about 20 months of age. A decrease or absence of tryptic activity has been observed in the majority of children sampled between 46 and 61 months of age which suggests that the development of the trypsin-inactivating microflora in humans is prolonged. Microbes capable of reducing cholesterol to coprostanol do not colonize the intestinal tract of humans until after the children are one year old. Bilirubin-metabolizing and mucin-degrading microbes are acquired by the majority of infants before the first birthday. Modification of mucins (glycoproteins in mucus) begins in the mouse colon from the time at which obligately anaerobic bacteria become established. Overall, these observations suggest that it is the obligately anaerobic inhabitants of the intestinal tract that are responsible for much of the modifications to host-derived molecules that occur in the intestinal habitat.

The acquisition of the normal microflora influences the type of short chain fatty acids detected in infant faeces. Acetic and propionic acid are commonly detected in the faeces of breast milk-fed infants, but isobutyric, butyric, isovaleric, valeric and isocaproic acids are present only after supplementary feeding begins, reflecting changes in the types of bacteria colonizing the intestinal tract as the biological succession proceeds. Supplemental feeding of breast milk-fed babies also appears to initiate a decrease in oxidation–reduction potential of the faeces. The E_h of the meconium is about $+175\,mV$, but that of the faeces from infants one to two days old is more reduced ($-113\,mV$). Adult values ($-348\,mV$) are not reached until after weaning.

The normal microflora contains species that are capable of causing disease under certain circumstances (see Chapter 5 for further discussion). The acquisition of the normal microflora, together with the fact that the immunological attributes of neonates are immature and lack prior exposure to specific pathogens, provide an explanation for the occurrence of certain infections in neonates. Acute purulent meningitis in human infants less than one month of age is commonly due to *Escherichia coli*. The involvement of this bacterial species in these infections is understandable when it is considered that not only can the infants be exposed to *E. coli* during passage through the vagina, but also that they harbour large populations of *E. coli* in their alimentary tract during the first weeks of life. The tissues of the infant are therefore exposed to large numbers of potentially pathogenic cells. Similarly, intestinal infections due to enterotoxigenic *E. coli* have a high incidence in neonatal farm animals. Mechanisms that maintain enterobacterial populations at low levels are not yet operating in the neonatal intestinal tract. Thus large populations of *E. coli* are present in the faeces of the animals, ensuring ease of transmission of infection between individuals if a pathogenic strain of *E. coli* should be present. *Candida albicans* is a member of the normal microflora of the human oral cavity that causes acute pseudomembranous candidiasis (oral thrush) in infants. In these cases, the lack of bacterial members of the microflora that normally suppress the replication of the yeast are lacking. Thus, coupled with the immunological immaturity of the infant, *Candida albicans* is able to proliferate and invades the oral mucosa.

FURTHER READING

Benno Y., Sawada K. and Mitsuoka, T. (1984) The intestinal microflora of infants: composition of fecal flora in breast-fed and bottle-fed infants. *Microbiology and Immunology* **28**, 975–86.

Cooperstock, M.S. and Zedd, A.J. (1983) Intestinal flora of infants, in *Human Intestinal Microflora in Health and Disease*, (ed. D.J. Hentges), Academic Press, New York, pp. 79–99.

Holdeman, L.V., Good, I.J. and Moore, W.E.C. (1976) Human fecal flora: variation in bacterial composition within individuals and a possible effect of emotional stress. *Applied and Environmental Microbiology* **31**, 359–75.

Moreau, M.-C., Thomasson, M., Ducluzeau, R. and Raibaud, P. (1986) Cinetique d'etablissement de la microflore digestive chez le nouveau-né humain en fonction de la nature du lait. *Reproduction Nutrition Developpement* **26**, 745–53.

Smith, H.W. (1961) The faecal bacterial flora of animals and man: its development in the young. *Journal of Pathology and Bacteriology* **82**, 53–66.

Stark, P.L. and Lee, A. (1982) The microbial ecology of the large bowel of breast-fed and formula-fed infants during the first year of life. *Journal of Medical Microbiology* **15**, 189–203.

Tannock, G.W., Fuller, R. and Pedersen, K. (1990) Lactobacillus succession in the piglet digestive tract demonstrated by plasmid profiling. *Applied and Environmental Microbiology* **56**, 1310–16.

Tissier, H. (1905) Repartition des microbes dans l'intestin du nourrisson. *Annales de l'Institut Pasteur (Paris)* **19**, 109–23.

Wood, B.J.B. (ed.) (1992) *The Lactic Acid Bacteria*, Vol. 1: *The Lactic Acid Bacteria in Health and Disease*, Elsevier Applied Science, London.

3 Sticky microbes:
the association of microbes with
host surfaces

Faced with the washout effect of host secretions (saliva, peristaltic movement of digestive tract contents), some members of the normal microflora adhere to, or associate with, epithelial or other surfaces in body habitats. Attached and replicating on a host surface, a microbial species can persist in a lotic (flowing) habitat whereas nonadherent microbes are washed away by the flow of secretion. Additionally, close association of microbes with host surfaces is likely to improve their access to potential nutrients that leak from epithelia. The pH at epithelial surfaces is generally close to neutrality and therefore may provide more suitable conditions for microbial growth in some instances than do the contents of, for example, the digestive tract. Oxygen diffusing from epithelia produces the microaerophilic conditions required by some mucosal surface-associated members of the normal microflora for growth. In the digestive tract, adherence of microbes to particles in the digesta permits the secretion of hydrolytic enzymes on to a substrate with which the microbial cells are associated, thus providing them with ready access to the products of hydrolysis (Figure 3.1).

3.1 SKIN

Most attention has been directed, in the case of the skin ecosystem, toward the adherence of *Staphylococcus aureus* to nasal epithelial cells because of the importance of asymptomatic staphylococcal carriage in the hospital environment (Section 1.2). Staphylococcal adherence (*in vitro*) to epithelial cells collected from infants on the first day of life is poor, perhaps due to the lack of appropriate receptors on the nasal cells. By five days after birth, however, staphylococcal adherence is similar to that obtained with epithelial cells collected from adult humans.

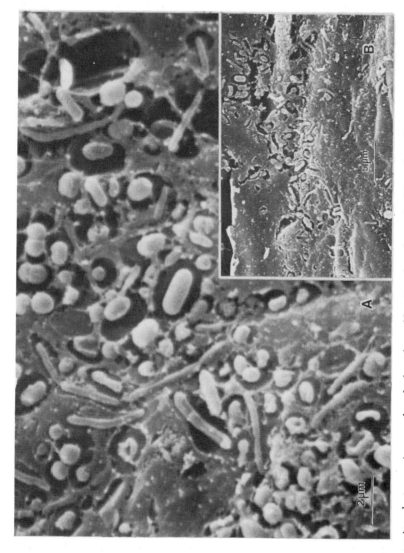

Figure 3.1 Scanning electron micrographs of wheat bran fibre in a faecal sample. Note the adherent bacterial cells in depressions that have resulted from digestion of the fibre by bacterial enzymes. Photographs provided by R.F. Adams.

3.2 ORAL CAVITY

The association of microbial cells with mucosal and dental surfaces in the oral cavity has been noted in Section 1.3. The adherence of oral bacteria to tooth surfaces has attracted considerable research interest because the formation of dental plaque is a prerequisite for dental decay. The following sequence of events has been observed when rigorously cleaned tooth surfaces are colonized by bacteria. Within a few minutes of completion of cleaning, streptococcal cells begin to associate with the tooth surface. These cells have either been recently displaced through the cleaning procedure or are carried from another oral site by the saliva. The bacteria do not attach directly to the enamel of the teeth: the tooth surfaces are coated by salivary glycoproteins which adsorb to the enamel surface. This covering layer on the teeth is called the acquired pellicle. Oral bacteria approaching the tooth surface face the same problems encountered by all microbes approaching a solid–liquid interface in any natural environment. Most surfaces, including those of bacterial cells, have an overall negative charge. Surfaces therefore attract a 'cloud' of positively charged ions which blanket them. As bacteria approach a surface they are first drawn towards it by attractive forces, but when closer to the surface repulsing forces are produced because of the similar charge on the bacterial surface and that which it approaches. The repulsing forces are sufficiently strong to prevent the bacterial cells making direct contact with the surface. Bacteria that adhere to surfaces synthesize extracellular materials or structures of very narrow diameter which can bridge the space between the bacterial cell and the surface adjacent. The bridging is possible because the repulsing forces are proportional to the diameter of the structure approaching the surface. Oral bacteria are initially held near the tooth surface by non-specific (ionic) forces. Specific, nonreversible adherence follows, apparently mediated by proteins on the bacterial cell surface. These adhesins interact with receptors located in the acquired pellicle that coats the tooth surfaces. Receptors include proline-rich proteins, α-amylase and possibly fragments of bacterial cell walls. Pioneering colonizers of the tooth surface include *Streptococcus sanguis*, *S. oralis*, *S. gordonii* and *S. mitis*. Other bacterial genera, while lacking adhesins that interact with acquired pellicle components, adhere specifically to bacterial cells already associated with tooth surfaces. Indeed, all oral bacterial types that have been investigated appropriately possess surface molecules that permit some sort of cell-to-cell interaction which can be intra- or intergeneric. Multiple adhesins are produced by some species so that they can participate simultaneously in associations with several different oral partners. *Actinomyces naeslundii*, for example, has adhesins that bind to receptors on the cell surfaces of *S. sanguis* and *S. gordonii*. *Veillonella atypica* adheres to streptococci and to *A. naeslundii*. *Prevotella loeschii* adheres to streptococci and *Actinomyces israelii*.

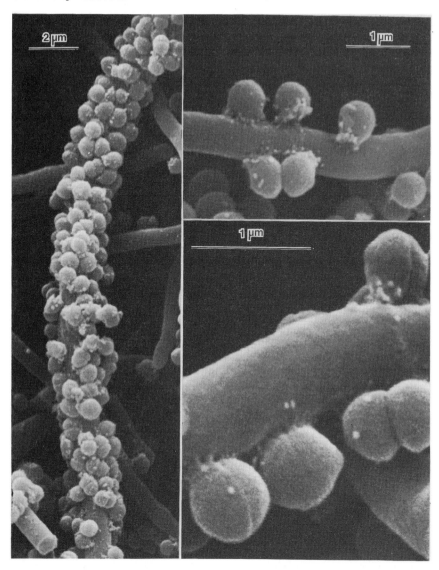

Figure 3.2 Scanning electron micrographs of coaggregating bacteria forming 'corncobs'. Filamentous microbe = *Corynebacterium matruchotii,* cocci = *Streptococcus sanguis*. Electron microscopy by R. Easingwood.

Additional bacterial species become associated with the developing plaque if it is permitted to accumulate. *Fusobacterium nucleatum* adheres to *A. naeslundii*, *V. atypica* and *A. israelii* among others and thus forms a bridge between the early and late colonizers of the tooth surface. Late colonizers that adhere to *F. nucleatum* include *Prevotella intermedia*, *Porphyromonas gingivalis* and *Selenomonas flueggei*. Interactions between the cells of *Corynebacterium matruchotii* and oral streptococci, as do those between *F. nucleatum* and other bacteria, give rise to 'corncob' configurations (Figure 3.2) in which a central rod-shaped cell is surrounded by adhering spherical cells. The attachment of the cells of different bacterial species to one another is referred to as coaggregation. Coaggregation is mediated by two classes of adhesin: proteins associated with the cell wall and fimbrial proteins. Some of the adhesins have carbohydrate-containing molecules as receptors and therefore belong to the category of proteins known as lectins. Aggregation phenomena in which lectins are involved are inhibited, *in vitro*, by the presence of specific carbohydrates (Table 3.1).

The adherence mechanisms of mutans streptococci (*S. mutans*, *S. cricetus*, *S. rattus*, *S. sobrinus*; Section 1.3) to tooth surfaces is of particular interest because these streptococci have been isolated from carious sites with a high frequency, and because they exhibit a pronounced ability to produce dental caries in experimental animals. The degree of cariogenic activity in the oral cavity is greatly influenced by the amount of sucrose ingested in the diet. The presence of mutans streptococci in the oral cavity is also related to the sucrose content of the diet. Specifically, high sucrose diets are associated with high cariogenic activity and a high incidence of mutans streptococci. Mutans streptococci synthesize two types of extracellular polysaccharide from sucrose. These polymers contain either glucose (glucans) or fructose (fructans) whose synthesis is catalysed by glucosyl transferases (GTFs) and fructosyl transferases (FTFs), respectively. Energy for the syntheses is derived from hydrolysis of the glycosidic bridge linking the glucosyl and fructosyl components of sucrose. This means

Table 3.1 Examples of coaggregation of oral bacteria

Bacterial species	Adhesin	Aggregation inhibited by
P. loeschii–S. oralis	Protein (75 kD)	Galactosides, especially N-acetylgalactosamine
P. loeschii–A. israelii	Protein (45 kD)	Not affected by mono- or oligosaccharides
P. gingivalis–F. nucleatum	Unknown	Galactosides
C. gingivalis–A. israelii	Protein (150 kD)	Neuraminic acid, neuraminlactose
F. nucleatum–various Gram-negative genera	Protein	Lactose
A. naeslundii–streptococci	Protein (fimbriae)	Galactosides

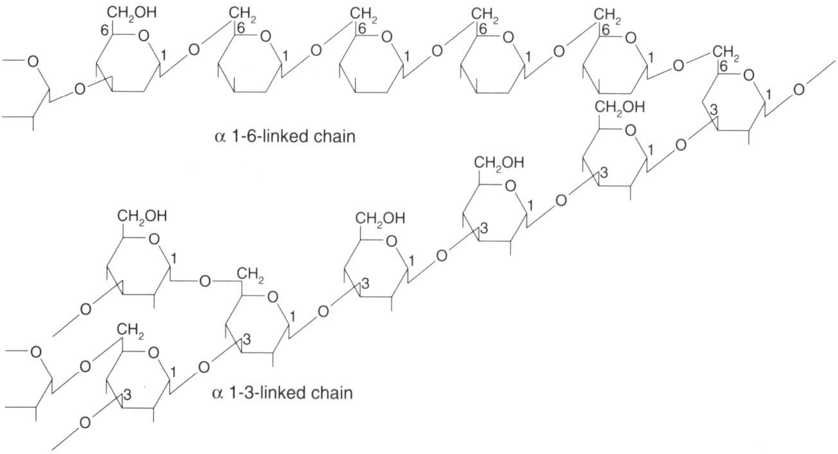

Figure 3.3 Generalized structure of a glucan synthesized by mutans streptococci. The branching polymer contains α-1-6- and α-1-3-linked regions.

that mutans streptococci cannot produce these polymers from individual hexose molecules. The glucans produced by mutans streptococci are of two types: water-insoluble glucans, also known as mutan, composed of highly branched glucose polymers in which α-1-3 glycosidic linkages predominate (Figure 3.3); and water-soluble glucans in which α-1-6 linkages predominate. The water-insoluble glucans are considered to be the most important in the colonization of tooth surfaces because they are cell surface-associated (bound to GTFs and to specific glucan-binding proteins), do not diffuse from plaque and cause aggregation of mutans cells. Further, they are not degraded by enzymes (dextranases, fructanase) produced by other oral microbes that remove water-soluble polymers from the habitat. The binding of water-insoluble glucans to mutans streptococcal surfaces ensures irreversible attachment to tooth and plaque surfaces.

3.3 GASTROINTESTINAL TRACT

Association of microbial cells with epithelial surfaces lining the gastro-intestinal tract has been observed commonly in studies of rodents (mice, rats), pigs and fowl. For example, in these species, lactobacilli are numerous in proximal regions of the tract because they can adhere to the surface of a stratified, squamous epithelium and multiply on it. An epithelium of this type is present in the forestomach of mice, the crop of fowl and the pars oesophagea of pigs (Figure 3.4). Adherence and replication of the bacteria results in the formation of a layer of bacterial cells on the epithelial

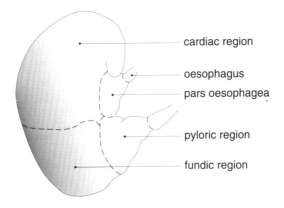

Figure 3.4 The regions of the porcine stomach.

surface (Figure 3.5). Lactobacilli shed from the layer provide a continual inoculation of the digesta and lactobacillus cells are therefore detected in large numbers throughout the gastrointestinal tract of these animal species (Table 3.2).

The mechanism by which lactobacilli adhere to epithelia has not yet been determined but preliminary *in vitro* investigations have shown that both carbohydrate and protein molecules are involved (Table 3.3). The molecular mechanism of attachment may be similar to that by which the cells of *Streptococcus pyogenes* adhere to pharyngeal cells: complexes of protein and lipoteichoic acid (LTA) at the streptococcal surface form fibrillae that bridge the gap between the bacterial and epithelial cell surfaces. The protein–lipoteichoic acid bridges are anchored at one end to the bacterial cell membrane and to fibronectin (a glycoprotein) in the epithelial cell membrane at the other via the hydrophobic (fatty acid) ends of lipoteichoic acid molecules (Figure 3.6). An alternative adherence mechanism is possible, however, involving proteins (lectins) on the lactobacillus cell surface that bind to specific carbohydrates. Indeed, the majority of mechanisms that permit attachment of bacteria to eukaryotic or microbial surfaces, and that have been studied in detail, utilize lectin–glycolipid or lectin–glycoprotein interactions. Lactobacillus adherence could be mediated by lectins synthesized by lactobacilli or host epithelial cells, or lectins derived from the animal's diet and which may coat epithelial cells (Figure 3.6). The nature of the receptors to which lactobacilli adhere on epithelial cells needs to be determined, because these bacteria exhibit host specificity. Strains originating in the rodent forestomach, for example, do not adhere to crop cells, while isolates from poultry do not adhere to epithelial cells from the rodent forestomach or porcine pars oesophagea.

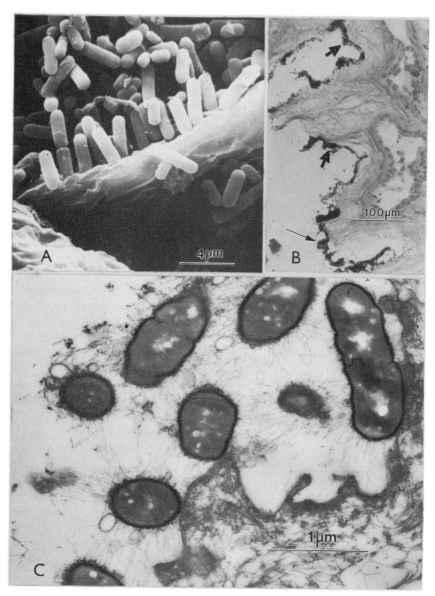

Figure 3.5 Lactobacillus cells adhering to epithelial surfaces. A: scanning electron micrograph of lactobacillus cells associated with the forestomach of a mouse; B: a layer (arrow) of lactobacillus cells associated with the forestomach epithelium of a mouse (Gram-stained section); C: transmission electron micrograph of lactobacillus cells adhering to an epithelial cell and to each other by means of fibrillae. Photograph A provided by D.C. Savage, reproduced by permission of the American Society for Microbiology; C provided by R. Fuller, reproduced by permission of the *Journal of Ultrastructure and Molecular Structure Research* (Academic Press).

Table 3.2 Lactobacillus populations detected in the digestive tract of mice, pigs, fowl and humans

Animal species	Stomach (crop^a)	Small bowel	Large bowel
Mouse	10^{8b}	10^7	10^8
Pig	10^9	10^7	10^9
Fowl	10^9	10^8	10^9
Human	$<10^4$	$<10^8$	$<10^9$

^{a}Fowl.
^{b}Approximate number of lactobacilli per gram of organ contents.

Table 3.3 Evidence from *in vitro* experiments that the adhesion of lactobacilli to epithelia requires protein and carbohydrate molecules

Adhesion to epithelial cells reduced after treatment of lactobacilli with	Molecules likely to be affected by the treatment
Heat (60 °C for 60 min or 100 °C for 10 min)	Proteins
Detergents (sodium dodecyl sulphate, Tween 80, Triton-X-100)	Proteins
Proteolytic enzymes (Papain, pepsin)	Proteins
Sodium periodate	Acid polysaccharide
Concanavalin A (monovalent)	Polysaccharides, glycolipids, glycoproteins containing α-D-glucose, α-D-mannose or β-D-fructose moieties

Sporing, segmented, filamentous bacteria (Figure 3.7) that have never been cultured in the laboratory but which can be detected in intestinal samples by microscopy, have been reported to occur in several vertebrate and invertebrate species (mouse, rat, dog, cat, monkey, sheep, poultry, termite and cockroach). Most studies of these organisms, however, have concerned mice and rats. The bacteria attach to the epithelium of the ileum and show a preference for cells associated with accumulations of lymphoid tissue (Peyer's patches) in the intestinal mucosa. The bacterial filament is attached by one end to the mouse tissue, the remainder of the filament extending into the lumen of the intestine (Figure 3.7). The attaching segment of the filament is inserted into a depression in the epithelial cell. The cytoplasm of the epithelial cells around these attachment sites contains polymerized actin, thus altering the cytoskeleton so that a permanent, 'socket'-like structure is formed around the bacterial segment. The relationship between the animal and the segmented, filamentous bacteria is

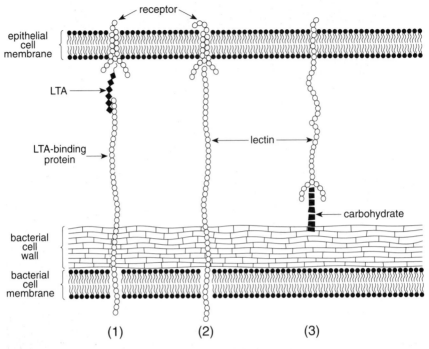

Figure 3.6 Possible mechanisms by which lactobacillus cells adhere to epithelial cells. (1) bacterial fibrillae composed of lipoteichoic acid (LTA) and protein molecules; (2) bacterial fibrillae composed of a lectin; (3) adherence mediated by a lectin on the epithelial cell surface.

complex, including a high degree of host specificity since strains harvested from mice do not colonize rats, and vice versa.

Lactobacilli and segmented, filamentous bacteria attach directly to epithelial cells. Some other intestinal bacteria associate with the layer of mucus covering the intestinal epithelium but do not actually attach to host cells. The bacteria colonize the mucus forming dense masses of bacterial cells close to the tissue surface. This phenomenon is best seen in preparations from the large bowel of mice and rats. In these animal species, the contents in the lumen of the large bowel contain a mixture of many hundreds of bacterial species. The mucus layer that covers the epithelium of the large bowel is colonized by fusiform-shaped (bacilli with tapered ends; Figure 3.8) and spiral-shaped bacteria (Figure 3.9). Crypts in the mucosa harbour, almost exclusively, spiral-shaped microbes such as *Helicobacter muridarum* and *Flexispira rappini* which are very motile in viscous gels such as that formed by mucus. At least some of the spiral-shaped bacteria associated with mucosal surfaces are microaerophilic rather than anaerobic.

Colonization of mucus associated with tissue surfaces by members of the normal microflora is very limited in humans compared to the situations

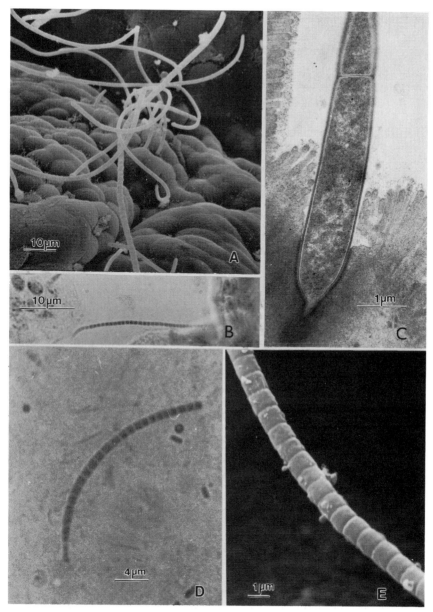

Figure 3.7 Segmented filamentous microbes adhering to the intestinal mucosa of mice. A: scanning electron micrograph showing the microbes adhering by one end of the filament to the murine mucosa; B: a filament observed in a Gram-stained section; C: transmission electron micrograph showing the site of attachment of a filament to a murine enterocyte; D: a filament in a Gram-stained preparation showing the tapered end of the filament which attaches to the intestinal mucosa; E: scanning electron micrograph showing the segmented nature of a filament. Photograph A provided by D.C. Savage; C provided by C.P. Davis; reproduced by permission of the American Society for Microbiology. Electron microscopy for E by R. Archibald.

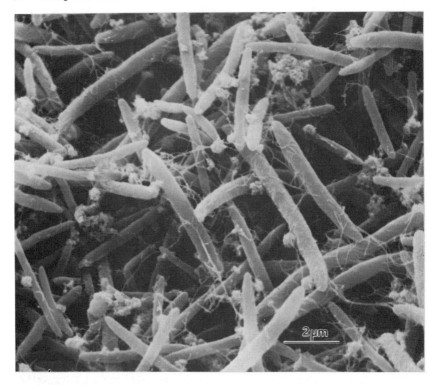

Figure 3.8 Scanning electron micrograph of fusiform-shaped bacteria that inhabit the large bowel of mice. Electron microscopy by C. Crichton.

described for rodents. Numbers of bacteria obtained from washed tissue surfaces from humans are considerably lower than those observed in studies of rodents. Epithelium-associated bacterial layers in human samples have only been demonstrated microscopically in the case of the rectum and anal canal where spiral-shaped microbes attached end-on to the tissue have been seen. A stratified, squamous epithelium is present only in the oesophagus and anal canal of humans, perhaps explaining why a large lactobacillus population is not characteristic of the gastrointestinal microflora of humans (Table 3.2).

FURTHER READING

Fuller, R. and Brooker, B.E. (1974) Lactobacilli which attach to the crop epithelium of the fowl. *The American Journal of Clinical Nutrition* **27**, 1305–12.

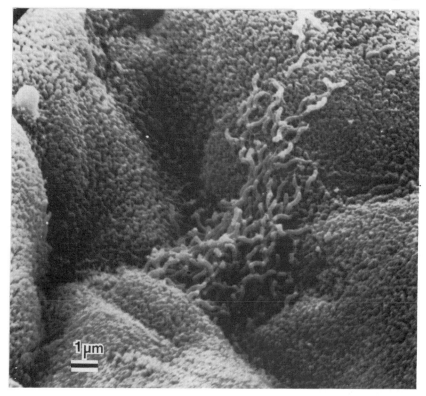

Figure 3.9 Spiral-shaped bacteria associated with the opening of a crypt in the large bowel mucosa of a mouse (scanning electron microscopy by C. Crichton).

Kolenbrander, P.E. (1991) Coaggregation: adherence in the human oral microbial ecosystem, in *Microbial Cell–Cell Interactions* (ed. M. Dworkin), American Society for Microbiology, Washington, D.C., pp. 303–29.

Savage, D.C. (1983) Morphological diversity among members of the gastrointestinal microflora. *International Review of Cytology* **82**, 305–34.

Slots, J. and Taubman, M.A. (eds) (1992) *Contemporary Oral Microbiology and Immunology*, Mosby Year Book, St Louis.

4 Invisible forces:
the influence of the normal
microflora on host characteristics

Silently, microbial communities inhabiting our body exploit the environments that we provide and ensure the continuance of their species. In exploiting their environment, the members of the normal microflora profoundly alter the biochemistry of the human body. This is apparent from the comparison of the characteristics of animals that harbour a normal microflora (conventional animals) with animals of the same species that lack a microflora (germfree animals).

Defined as animals free of all demonstrable forms of microbial life, germfree (axenic) animals are essentially 'pure cultures' of a particular animal species. As is often the case in microbiology, the concept of rearing animals free of microbial associates was first expressed by Louis Pasteur. After presenting the work of his student Duclaux on nitrogen fixation with sterile or inoculated plants, Pasteur expressed his views to the French Academy of Science:

> In presenting this work of Duclaux, I take the liberty to offer an idea for an experiment which comes not only from the evidence which I give to the Academy in his name, but also from work, no less distinguished, which he has already done on the role of bacteria in digestion.
>
> For several years during discussions with young scientists in my laboratory, I have spoken of an interest in feeding a young animal (rabbit, guinea pig, dog or chicken) from birth with pure nutritive material. By this expression I mean nutritive products which have been artificially and totally deprived of the common microorganisms.
>
> Without affirming anything, I do not conceal the fact that if I had the time, I would undertake such a study, with the preconceived idea that under these conditions, life would be impossible.
>
> If this work could be developed simply, one should be able to study digestion by the systematic addition to the pure food, of one or another single microorganism or diverse microorganisms with well defined relationships.

A chicken egg could be used without serious difficulty for this type of experiment. Before the chick is hatched the exterior could be cleaned of all living organisms; then the chick would be placed in a cage without any kind of microorganisms. In this cage where pure air is given, the chick would also be supplied with sterile food (water, milk, and grains).

Whether the result would be positive and confirm my preconceived view, or whether it would be negative, in other words, that life would be easier and more active, it would be most interesting to perform the experiment.

Louis Pasteur (1885)

Although Pasteur did not pursue his ideas regarding the derivation of germfree animals, success in doing so was eventually achieved by others. Germfree vertebrates (guinea pigs) were first derived in 1895 by Nuttall and Thierfelder. By 1962 germfree rodents had become commercially available in the USA and research facilities rearing several species of germfree animals had been established in Europe and Japan. Most germfree research in recent times has been conducted using piglets, chickens, mice and rats. Germfree mammals are derived by aseptic caesarian delivery of young animals from their conventional mother. The foetus in the uterus is microbiologically sterile (apart from those viruses which can cross the placenta). The young animal is placed directly into a sterile container (called an 'isolator'), usually made of flexible plastic, which is supplied with a filtered air supply. The young animal is hand-fed sterile milk of an appropriate formulation until it is able to ingest sterile solid food. Germfree chicks are derived by sterilizing the surface of freshly-laid fertile eggs, incubating them in a disinfected incubator, and introducing the eggs into an incubator in a sterile isolator shortly before hatching. The germfree animals are maintained in the isolator (Figure 4.1), being supplied with sterilized water, food, cages and bedding. Introduction and removal of material or animals from the isolator is possible because of a double door system and sleeves with gloves set into one wall of the apparatus. Germfree rodent colonies can be established once the animals are old enough to breed, and further colonies can be started in other locations using a few pairs of mice or rats transported to the new site to form the nucleus of another colony. Germfree chickens and germfree mammals of larger species are derived as required for each experiment.

The characteristics of ex-germfree animals colonized with one or several types of microbe can be compared with those of noncolonized controls. Changes (microflora-associated characteristics) produced in the host animal by the microbial associates are then apparent. Germfree animals and those which have been colonized with specified microbial strains are called gnotobiotic (known life) animals. Comparisons of the characteristics of germfree and conventional animals have revealed the extent to which the

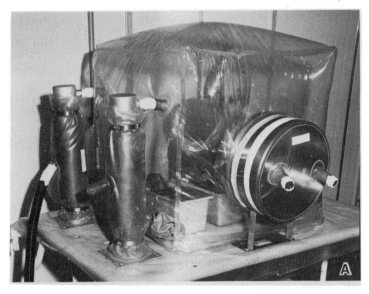

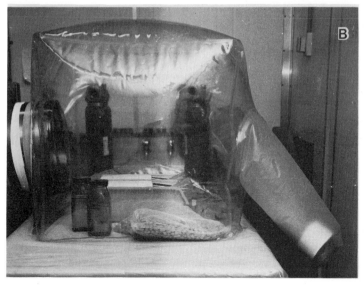

Figure 4.1 An isolator used in gnotobiotic research. Note the air filters in photograph A and the sleeves and gloves in B.

biochemistry and physiology of the host are influenced by the normal microflora (Table 4.1).

Care is needed in the planning and interpretation of the results of gnotobiotic experiments. Germfree animals are rather elegant test tubes which provide host factors which cannot be reproduced in the laboratory.

Table 4.1 Some differences between germfree and conventional rodents

Characteristic	Germfree compared to conventional
Basal metabolic rate	Lower
Cardiac output	Less
Blood volume	Less
Mucosal surface area in small bowel	Less
Mucosal cell turnover rate in small bowel	Slower
Intracellular epithelial cell enzyme concentrations	Higher
Lamina propria	Thinner
Reticuloendothelial elements in small bowel wall	Less
Small bowel motility	Slower
Wet weight of caecum with contents	Greater
Urea in caecal contents	Present (none detected in conventional)
Ammonia in caecal contents	Less
Ammonia concentration in portal blood	Less
Ileocaecal lymph nodes	Smaller
Bile acid transformation in intestine	None (secondary bile acids in conventional)
Serum immunoglobulins	Less (some fractions absent)
Trypsin activity in large bowel contents	Present (inactivated in conventional)
Urobilinogens	Absent in germfree
β-aspartyl-glycine	Present (degraded in conventional)
Coprostanol	Absent in germfree
Short chain fatty acids (propionic, butyric)	Absent in germfree
β-glucuronidase	Absent in germfree

The ways that microbes behave in association with germfree animals, however, are often influenced by the nature of the experiment. The order in which bacterial strains are introduced into the gastrointestinal tract of germfree rodents will sometimes influence which microbe becomes numerically dominant. A single microbial type colonizing the gastrointestinal tract of a gnotobiote usually attains a much higher population level than it does in a conventional animal where the microbe is faced with intense competition from the other members of the normal microflora. Physiological differences between germfree and conventional animals can also influence microbial colonization patterns. A single bacterial strain colonizing the gastrointestinal tract of a gnotobiote, for example, will usually be present in large numbers in both the small and large bowel of the animal. In conventional animals the microbe may be restricted to the large bowel because of the swift flow of intestinal contents through the small bowel. Small bowel motility is slower in gnotobiotic animals and microbial colonization of that region can occur more easily.

Relating the results obtained from gnotobiotic experiments to 'real life' situations is another problem encountered in this type of research. The relevance of associating germfree mice with microbes isolated from humans is debatable, especially since we know that some members of the normal microflora show specificity for particular animal hosts. Lactobacilli which colonize epithelial surfaces in the gastrointestinal tract, for example, show specificity for either rodents, pigs or chickens (Section 3.3). Gnotobiotic research, together with medical microbiology, rumen microbiology and nutrition research have, however, contributed to our understanding of the ways in which the normal microflora influences our bodies.

4.1 NUTRITION

Ruminants rely on the normal microflora of the gastrointestinal tract for the digestion of plant-derived substances in their diet and for the provision of essential amino acids and vitamins. Plant structural materials (cellulose, hemicelluloses) form the major components of the diet of ruminants such as sheep, cattle, goats and deer. The ruminants are unable to synthesize digestive enzymes capable of degrading these plant substances. The ruminant gastric region is formed into four compartments: the rumen, reticulum, omasum and abomasum. The rumen–reticulum (Figure 4.2)

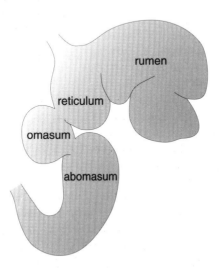

Figure 4.2 The compartments that comprise the proximal region of the ruminant gastrointestinal tract.

accommodates a large collection of microbes capable of degrading the plant material. The rumen–reticulum forms 85% of the total stomach storage capacity, the digesta contained within these organs forming up to 20% of the animal's body weight. The digesta remains in the rumen–reticulum until about 80% of the total digestable dry matter has been fermented by microbial action. Rumen liquor containing microbial cells and dietary residues passes from the rumen–reticulum into the omasum which is an absorptive organ, and thence into the abomasum which has a secretory mucosa that produces acid gastric juice containing pepsin. The abomasum is thus the equivalent of the stomach in monogastric animals such as humans. The compartmented gastric region of ruminants permits a semi-continuous fermentation to occur simultaneously with, and separately from, a subsequent peptic digestion of the food.

The rumen–reticulum ecosystem is very anaerobic ($E_h = -400\,\text{mV}$) and harbours a complex collection of mostly nonsporing anaerobic bacteria (10^{10} per gram) and anaerobic protozoa (10^6 per gram), as well as anaerobic fungi. These latter microbes, which belong to the genus *Neocallimastix*, are particularly important in the initial degradation of tough plant fibre material (Figure 4.3). The protozoa are not essential to the fermentation process (they can be removed by chemical treatment without impairing ruminant digestion) but they are predators of certain bacterial types and thus regulate bacterial numbers. Fifteen genera of anaerobic bacteria and eight genera of protozoa have been detected in the rumen. Some commonly encountered rumen microbes are listed in Table 4.2. The rumen fermentation first involves the breakdown of plant structural polymers (e.g. cellulose) by some of the rumen bacteria which possess the appropriate enzymic capability (e.g. cellulase activity). The carbohydrate residues released by these initial reactions are available for fermentation by these and other microbial types. The overall rumen fermentation is described by the equation

$$57.7 \text{ moles plant carbohydrate} \rightarrow 65 \text{ acetic} + 20 \text{ propionic} + 15 \text{ butyric}$$

$$+ 60 \text{ } CO_2 + 35 \text{ } CH_4 + 25 \text{ } H_2O$$

The short chain fatty acids (acetic, propionic and butyric) resulting from the rumen fermentation are absorbed from the rumen–reticulum into the bloodstream and provide the main source of energy for the ruminant. Plant proteins in the diet are also degraded by microbial action. Some of the resulting peptides and amino acids are converted to proteins in microbial cells, but some are deaminated and others fermented to short chain fatty acids. Nearly all of the rumen microbes can utilize ammonia as a source of nitrogen. Plant lipids are degraded by the microbes and glycerol is fermented to propionic acid. Unsaturated fatty acids are hydrogenated. Rumen bacteria contain odd-numbered carbon straight- and branched-chain

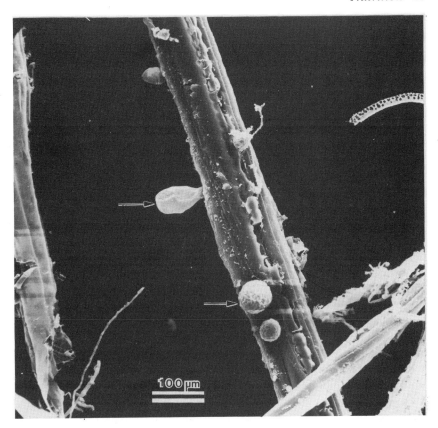

Figure 4.3 Scanning electron micrograph of rumen contents showing a plant stalk to which is attached sporangia (arrows) of *Neocallimastix*. Electron microscopy by R. Archibald.

saturated fatty acids. Rumen microbes carried out of the rumen–reticulum into the abomasum and small bowel are digested by mammalian processes and thus serve to supply the ruminant with most of its requirements for amino acids, vitamins and lipids. As one reflection of the reliance of ruminants on microbes for nutrients, ruminant fat is rich in odd-numbered saturated fatty acids.

The degradative processes occurring in the rumen–reticulum are the result of the metabolic interactions of numerous microbial species. Some of the rumen bacteria in pure culture produce fermentation products which cannot be detected in the rumen itself. This is because the metabolic products of some microbes (termed rumen intermediates) become the substrates for the metabolic activities of yet other microbial types. Hydrogen gas, lactic acid and succinic acids belong to this category (Figure 4.4).

Table 4.2 Some commonly isolated rumen microbes

Fibrobacter succinogenes	Resembles *Bacteroides*; cellulolytic.
Butyrivibrio fibrisolvens	Gram-negative curved rods motile by polar or subpolar flagella. Obligate anaerobe. Butyric acid is a major fermentation product.
Clostridia	Sporing bacilli. Obligate anaerobe.
Ruminococcus albus	Gram-positive cocci; obligate anaerobe; cellulolytic; acetic and formic acids, hydrogen and carbon dioxide as fermentation products.
Streptococcus bovis	Gram-positive cocci; starch decomposer; lactic acid as fermentation product.
Succinimonas	Gram-negative rods motile by a single polar flagellum. Obligate anaerobe. Succinate and acetate as the main fermentation products.
Selenomonas	Gram-negative curved to helical rods. Motile with active tumbling. Up to 16 flagella are arranged linearly as a tuft near the centre of the concave side. Obligate anaerobe. Acetic and propionic acids, carbon dioxide and/or lactate as fermentation products.
Succinivibrio	Gram-negative curved, helically twisted, rods with pointed ends. Motile by single polar flagellum. Obligate anaerobe. Succinic, acetic, formic and sometimes lactic acids as major fermentation products.
Veillonella	Gram-negative cocci. Obtain energy from fermenting pyruvic and lactic acids. End products of fermentation are acetic and propionic acids, carbon dioxide and hydrogen. Obligate anaerobe.
Peptostreptococci	Gram-positive cocci. Obligate anaerobe. Can metabolize peptone and amino acids.
Lachnospira multiparus	Weakly Gram-positive long curved rods. Monotrichous lateral to subpolar flagella. Obligate anaerobe. Pectin decomposer. Formic, acetic and lactic acids, ethanol, carbon dioxide and hydrogen as fermentation products.
Desulfovibrio	Gram-negative curved to straight rods, motile by polar flagella. Obligate anaerobe. Anaerobic respiration with sulphate or other sulphur compounds as terminal electron acceptors, being reduced to hydrogen sulphide.
Lactobacilli	Gram-positive rods that produce lactic acid as the major fermentation product. Grow best under anaerobic conditions.

Table 4.2 *Continued*

Methanobrevibacter ruminantium	Gram-positive rods; methanogenic. Obligate anaerobe.
Methanomicrobium mobile	Gram-negative rods; motile; methanogenic. Obligate anaerobe.
Ciliated protozoa	Holotrichs: completely and uniformly ciliated. *Isotricha* and *Dasytricha*. Utilize soluble carbohydrates. Entodiniomorphs: a zone of adoral and (usually) a dorsal zone of cilia. *Diplodinium* and *Entodinium*. Ingest particulate matter.

Methane-producing bacteria perform an important role in the rumen by utilizing carbon dioxide and hydrogen produced by fermentative bacteria. The methane bacteria maintain a low partial pressure of hydrogen in the rumen ecosystem which permits fermentative bacteria to oxidize NADH by a more energetically favourable process than would be the case if the gas accumulated in the ecosystem. The methane bacteria's utilization of hydrogen is an example of interspecies hydrogen transfer and affects the type and amount of end-products produced by the fermentative bacteria.

The contribution that the members of the normal microflora of the large bowel make towards host nutrition is less clear than in the case of the rumen. Monogastric animals and birds with well-developed caeca harbour a collection of microbes capable of degrading complex plant materials. Microbial activity in the large bowel of rats, for example, degrades 39%

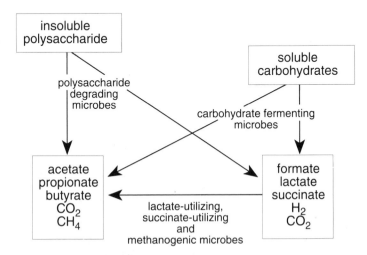

Figure 4.4 The microbial digestion of dietary carbohydrates in the rumen.

of dietary cellulose. Horses possess a large caecum (25–30 l capacity) and colon (55–70 l) in which up to 50% of grass cellulose is degraded by microbial activity. Even in the pig, which has a relatively small caecum, 40–70% of ingested cellulose is digested in the large bowel. The caecum (Figure 4.5) is a thin-walled, highly vascularized organ. Mixing of caecal contents occurs by muscular contractions and a relatively large absorptive surface area is available. Short chain fatty acids are present in the large bowel contents. The acids enter the bloodstream of the host, but their contribution to daily energy requirements, unlike the situation in ruminants, may be minimal. Cellulose fermentation in the large bowel is thought to contribute 10–12% of daily energy requirements for the rabbit, up to 5% for rats, and no more than 2% for pigs and humans. Vitamins are synthesized in the large bowel by members of the normal microflora. Facultatively anaerobic bacteria such as *Escherichia coli* are able to synthesize a wide range of vitamins including biotin, riboflavin, pantothenic acid, pyridoxine and vitamin K. These and other vitamins can be detected in the large bowel contents of animals and birds. Germfree rats fed a diet deficient in vitamin K develop the haemorrhagic syndrome associated with such a deficiency. Conventional rats fed the same diet do not become diseased. Germfree rats inoculated with caecal contents from conventional rats do not suffer from the deficiency disease when fed the vitamin K-deficient diet. Similarly, inoculating germfree rats with vitamin K-synthesizing *E. coli* removes the deficiency. Some animal species therefore do obtain nutritional benefit from the normal microflora of the large bowel. Vitamins and most of the other constituents of microbial cells present in the large bowel can only be utilized by monogastric animals, however, if the bowel contents can be recycled through the nutrient-absorbing region of the digestive tract: the small bowel. Some animals (e.g. rodents and rabbits) consume a large proportion of their daily faecal output in the form

caecum

colon

stomach   small bowel

Figure 4.5 The rodent gastrointestinal tract, showing the location of the caecum.

of soft pellets eaten directly from the anus. The pellets then pass through the digestive tract and the nutrients contained within the faecal material become available to the host's tissues. This habit, known as coprophagy, is beneficial to the host: conventional rats prevented from indulging in coprophagy do not thrive unless biotin and pantothenic acid are added to the diet. Coprophagy, literally and nutritionally, can be said to be 'making ends meet'.

It is likely that monogastric mammals and gallinaceous birds in the feral state obtain more benefit from harbouring a microflora in their large bowel than do domesticated or laboratory-raised animals. Feral animals may often be unable to fully satisfy their nutritional requirements from dietary sources because of competition for food from other animals or because of climatic or other natural conditions. The nutritional contribution of the normal microflora under these circumstances is probably critical to the health of the host. In some animals and gallinaceous birds, the caecum makes up as much as 5% of the animal's body weight. The retention of such a large organ by these animals must be beneficial to the host because added body weight is detrimental in terms of energy expenditure for a speedy escape from predators.

Animals that do not indulge in coprophagy apparently do not benefit to the same extent as rats and rabbits from the metabolism of their intestinal microbes. Vegetarian humans, for example, can develop vitamin B_{12} deficiency because of a deficient diet, even though bacterial production of vitamin B_{12} occurs in their colon.

The microbes that inhabit the large bowel of humans, like those inhabiting the distal regions of the digestive tract in other animals, encounter diet-derived substrates that the host has been unable to digest. It has been calculated that about 20 g of plant structural material is ingested per day by individuals consuming a 'western' diet. Five to 10 g of dietary fibre can be recovered in faeces. Since humans do not have enzymes for the digestion of plant structural substances, microbial activity must have digested the remainder. About 50% of ingested cellulose and 70–90% of hemicelluloses are fermented in the large bowel of humans. Ten to 15% of starch in foods such as oats, white bread and potatoes escapes digestion in the small bowel and is fermented by microbes in the large bowel. Most of this fermentative activity occurs in the right colon (Figure 4.6). Plant-derived carbohydrates are not the only substrates available for microbial nutrition: intestinal secretions and cells form a substantial source of fermentable material. Intestinal glycoproteins, for example, are 80% carbohydrate. A list of the major polysaccharide-utilizing bacteria in the human colon is provided in Table 4.3. If 30 g of carbohydrate were fermented daily in the colon, 200 mmol of fermentation products (mostly short chain fatty acids) would be produced. Five to 20 mmol of fatty acids (acetic, propionic and butyric acids) are actually detected in faeces, which means that most

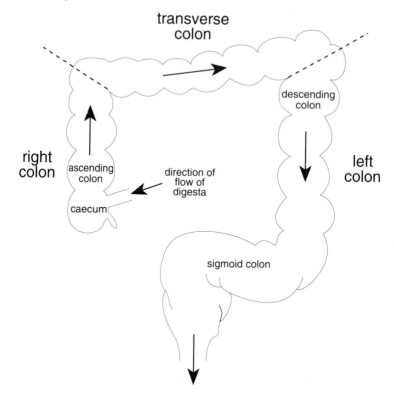

Figure 4.6 The human large bowel.

of the microbial fermentation products are absorbed by the host, or are degraded by colonic bacteria. Short chain fatty acids produced in the colon are not thought to contribute significantly to the daily energy requirements of humans. However, laboratory experiments involving the measurement of oxygen uptake by colonocytes in relation to various substrates indicate that butyric acid is the predominant energy source for the epithelium of the human colon. A proportion of hydrogen and carbon dioxide produced in the large bowel during fermentation of carbohydrates is converted to methane (0.03–3 l per day) in up to 50% of the human population. However, in contrast to the situation in the rumen, methane production does not appear to influence short chain fatty acid synthesis by the fermentative bacteria inhabiting the large bowel of humans.

Table 4.3 Examples of polysaccharide-utilizing bacteria in the human colon

Polysaccharide	Some strains of
Cellulose	Bacteroides
	Ruminococcus
Non-cellulosic	Bacteroides
β-glucans	Peptostreptococcus
Arabinogalactan	Bacteroides
	Bifidobacterium
Xylan	Bacteroides
	Bifidobacterium
	Ruminococcus
Plant gums (e.g. guar gum)	Bacteroides
	Bifidobacterium
	Ruminococcus
Pectin	Bacteroides
	Eubacterium
Alginate	Bacteroides
Mucopolysaccharides	Bacteroides
Mucin	Bacteroides
	Ruminococcus
	Bifidobacterium

4.2 ENTEROHEPATIC CIRCULATION

The diet contains many substances that are not of nutritional value: plant products, food additives, drugs and environmental contaminants, both natural and synthetic. These foreign compounds (xenobiotics) may be absorbed from the small bowel. Some of the compounds are potentially toxic to the host and are detoxified by the liver, having travelled to this organ in the blood by the portal vein. Detoxification by the liver is often a two-stage process: oxidation to neutralize active groups, then conjugation to form glucuronides or sulphates to make the molecule more polar and thus facilitate excretion. Most foreign compounds are excreted in the urine, but smaller molecules are efficiently excreted in the bile. Unabsorbed xenobiotics and conjugated molecules that enter the intestine in bile are available for metabolism by microbes. Deconjugation of molecules by microbial metabolism can set up an enterohepatic circulation in which molecules are transported between the intestinal lumen and the liver via the portal blood circulation and bile (Figure 4.7). Not only the metabolism of xenobiotics is influenced by microbial activities: the excretion of host-derived molecules is also affected. The excretion of bilirubin, a product of haemoglobin catabolism, for example, is markedly influenced by an enterohepatic circulation in which intestinal microbes are involved.

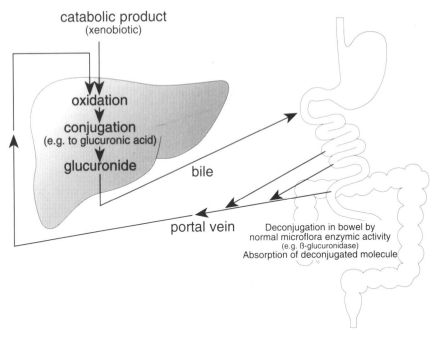

Figure 4.7 The enterohepatic circulation of host-derived (i.e. catabolic products) or exogenously-derived (xenobiotic) molecules.

Approximately 90% of bilirubin entering the liver is linked to glucuronic acid molecules in the liver to form a diglucuronide (the remainder is conjugated with sulphate). The diglucuronide molecules pass to the bowel in the bile and about 5% is deconjugated by bacterial β-glucuronidase. Free (deconjugated) bilirubin is absorbed from the bowel and passes back to the liver by the portal blood circulation. The remaining conjugated bilirubin is reduced by bacterial activity to form urobilinogens, some of which are excreted in the faeces, while the rest is absorbed into the blood and excreted via the kidneys. Enterohepatic circulation of a substance, by cycling it between bowel lumen and body tissues, prolongs its half-life in the body, and is therefore important in the pharmacological action of some drugs (e.g. chloramphenicol and morphine). Concurrent antibiotic therapy administered by mouth is one of the commonest factors associated with contraceptive pill 'failure' (i.e. conception occurs). This may be due to the antibiotic reducing the populations of intestinal microbes that normally deconjugate the oestrogens/progestegens present in contraceptive drugs. Without deconjugation, the drugs are excreted in the faeces rather than being reabsorbed into the blood stream at a sufficiently high concentration to mimic pregnancy and hence prevent ovulation and implantation.

4.3 MICROBIAL INTERFERENCE

This is the phenomenon by which microbes already established in an ecosystem prevent other microbial types from colonizing that site. Although microbial interference occurs throughout nature, it is of particular interest in the case of ecosystems associated with animal bodies. Microbial interference, also known as colonization resistance, operates in animal-associated ecosystems as a first line of defence against infection by pathogenic microbes that cause communicable diseases. To gain access to the tissues of the host, the pathogen must often first survive in an environment already colonized by microbes that are extremely well-adapted for life in that ecosystem. The phenomenon of microbial interference exerted by the normal microflora is therefore an important nonspecific resistance mechanism of the host animal.

Microbial interference has been most thoroughly investigated in relation to the gastrointestinal tract. Evidence that microbial interference is an important resistance mechanism in the gastrointestinal tract is provided from three types of study: antibiotic-treated subjects, gnotobiotic animals and 'stressed' animals.

4.3.1 Antibiotic-treated individuals

The use of broad-spectrum antibiotics such as chlortetracycline and oxytetracycline in the 1950s drew attention to the fact that the normal microflora helps to prevent the establishment of pathogens in that region of the host animal. A proportion of human patients treated orally with these antibiotics developed colitis, apparently, but not convincingly, demonstrated to be due to colonization of the intestinal tract by *Staphylococcus aureus* as a result of disruption to the normal microflora. More recently, the association of toxin-producing *Clostridium difficile* strains with colitis in antibiotic-treated hospital patients has provided evidence of the importance of microbial interference. Pseudomembranous colitis, associated with the oral administration of ampicillin, clindamycin, cephalothin or other antibiotics, is characterized by watery, usually nonbloody diarrhoea and abdominal pain. Sigmoidoscopy demonstrates white–yellow areas (plaques) on the surface of the colonic mucosa which, in severe cases, coalesce to form a sheet composed of fibrin, mucins and leucocytes (a pseudomembrane) on the epithelial surface. The disease can contribute directly to patient death. The signs and symptoms of pseudomembranous colitis and in many cases of the less severe disease, antibiotic-associated colitis, are due to the proliferation of an obligately anaerobic bacterium, *Clostridium difficile* in the large bowel. Attaining populations of about 10^6 per gram of large bowel contents, the clostridia secrete two toxins: toxin A and toxin B. Toxin A is an enterotoxin since it causes fluid accumulation in ileal loops formed experimentally in rabbits. The toxin (308 kD protein) also has cytotoxic activity since it causes a cytopathic effect on cultured mammalian cells. Toxin B (270 kD protein) is a cytotoxin and is lethal to experimental animals when administered with a sublethal dose of toxin A. It is proposed that toxins A and B act synergistically, the former damaging the epithelium and thus allowing toxin B access to underlying tissue. That both toxins are important in the aetiology of disease is demonstrated by the observation that hamsters require immunization against both toxins to be protected from ileocaecitis caused by *C. difficile*.

The proliferation of *C. difficile* in the large bowel results from the suppression of other members of the normal microflora during prolonged antibiotic therapy (Table 4.4). The absence of these members of the

Table 4.4 Antibiotics and the intestinal microflora

Antibiotics having a major effect on the colonic microflora
(*Bacteroides*, anaerobic cocci, bifidobacteria, lactobacilli eliminated or decreased in numbers; streptococci increased)
Ampicillin
Cefoperazone
Clindamycin
Neomycin + tetracycline, erythromycin or metronidazole
Kanamycin + tetracycline, erythromycin or metronidazole

Antibiotics having a moderate effect on the colonic microflora
(Coliforms and obligate anaerobes slightly decreased in numbers; streptococci slightly increased)
Neomycin
Kanamycin
Cotrimoxazole
Cephalin
Cefoxitin
Moxalactam
Erythromycin

Antibiotics having minor effects on the colonic microflora
Sulphonamides
Penicillins G and V
Bacampicillin
Dicloxacillin
Cefotaxime
Imipenem
Tetracyclines (major effect soon after its introduction, but nowadays many bacteria are resistant)
Chloramphenicol
Colistin
Metronidazole

(Source: S.M. Finegold (1986) Intestinal microbial changes and disease as a result of antimicrobial use. *Pediatric Infectious Disease* 5, S88–90.)

microflora removes from the ecosystem the mechanisms by which *C. difficile* populations are normally regulated. In the antibiotic-altered ecosystem, *C. difficile* replicates to levels where its toxins are at pathological concentrations. *Clostridium difficile* is present in the faeces of about 3% of healthy subjects but is easily transmitted between patients in hospital wards. Results from *in vitro* investigations using faecal homogenates inoculated with *C. difficile* point to the involvement of obligately anaerobic bacteria as the members of the microflora that normally regulate the population of *C. difficile*. The precise species that achieve this effect have not, however, been identified.

Microbial interference is also of consequence in the treatment of humans with granulocytopenic (paucity of white blood cells) disease, or receiving immunosuppressive therapy for malignancy. Such patients are much more likely to contract infections than normal because their antimicrobial defences are not operating optimally. Various antibiotic regimens have been used to 'decontaminate' the oropharynx and gastrointestinal tract of these patients because their infections are often due to microbes (such as *E. coli*, yeasts and *Pseudomonas* species) that inhabit these regions. Some broad-spectrum antibiotics may actually worsen the situation, however, because by disrupting the normal microflora, they permit the replication of *Enterobacteriaceae*, *Pseudomonas* and *Candida* which increase in numbers because the anaerobes that usually suppress their multiplication have been killed by the antimicrobial drugs. The enterobacteria, pseudomonads and yeasts are the very organisms likely to produce infection in the immunologically compromised patients. Van der Waaij has recommended that such patients be treated with antibiotics which are known to leave the anaerobic component of the normal microflora, as well as Gram-positive bacteria such as streptococci, staphylococci and enterococci, intact while inhibiting the potentially pathogenic microbes (selective decontamination). A combination of antibiotics, one of which is present systemically and the others locally is recommended for this purpose: cefotaxime (systemic), polymyxin and tobramycin or norfloxacin, or neomycin or gentamicin (local), and amphotericin B or nystatin (local) have been used in this context. Nearly all studies that have investigated the effectiveness of selective decontamination in intensive care facilities have demonstrated a decreased incidence of lower respiratory tract infection in treated patients compared to controls. The studies do not show, however, reductions in the cost of hospital care required per patient, reduced length of stay in intensive care, days of mechanical ventilation required, or mortality.

4.3.2 Gnotobiotic animals

Germfree animals are more readily colonized and infected by gastrointestinal pathogens than are conventional animals. As few as 10 viable

Salmonella enteritidis cells, for example, introduced into the gastro-intestinal tract of germfree mice result in the death of all challenged animals within 5–8 days. The corresponding value for challenged conventional mice is about 10^9 salmonella cells. The ease with which salmonella can establish in the intestinal tract of germfree mice is apparent from the fact that, three days after inoculation with 10^4 *S. typhimurium* cells intragastrically, germfree mice have population levels of 10^7–10^8 salmonella per gram of intestine. In contrast, germfree mice inoculated with caecal contents from a conventional mouse prior to challenge with salmonella have few, if any, salmonella in the intestinal tract.

4.3.3 Stressed animals

Domestic animals, such as sheep, cattle and pigs, maintained under harsh environmental or dietary conditions are known to be more susceptible to infection by gastrointestinal pathogens. Holding animals in pens or yards under normal farm management practices, transportation or radical changes in diet can lead to outbreaks of salmonella infection. The 'precipitating' or 'stress' factors which precede such outbreaks no doubt alter the animal's physiology in ways which render their tissues more susceptible to infection. The normal microflora of stressed animals is also markedly altered, and the breakdown of regulatory mechanisms in the gastro-intestinal ecosystem allows the easier establishment of pathogens in the tract.

The mechanisms which mediate microbial interference are not well-studied, but some factors that are believed to be important are listed in Table 4.5. It is probable that some interference mechanisms are dominant under certain circumstances but are replaced by other mechanisms when conditions change. Thus competition for a carbon source could be replaced by inhibition due to the production of toxic substances when the availability of the carbon source is no longer a limiting factor. There is therefore a degree of redundancy in microbial interference: several mechanisms may exist which will regulate a given microbial population, but each may operate only under certain conditions.

Studies of microbial interference affecting resistance to infection with *Salmonella* species stretch back over 40 years. This is therefore one of the better studied areas concerning microbial interference. The results of these studies allow us to conclude that the normal microflora contributes to host resistance to salmonella infection, at least in mice, in the following ways.

1. Certain members of the normal microflora influence physiological factors in the small bowel (e.g. the rate at which the digesta is moved through the intestinal tract), which in turn make it more difficult for salmonella to establish in that site.

Table 4.5 General mechanisms of microbial interference

Competition Two or more microbial types in rivalry for a factor in the environment which is not present in sufficient quantity to satisfy the demands of all of the organisms (competition for nutrients, competition for space on surfaces).

Amensalism The inhibition of the activities of one or more microbial types by the production of a toxic substance by another type (e.g. hydrogen sulphide, antibiotic-like substances (bacteriocins), lactic acid, short chain fatty acids, transformed bile acids).

Parasitism One microbial type consumes another larger type. The importance for microbial interference of the parasitism of microbes by viruses and small parasitic bacteria (bdellovibrios) is not known.

Predation One microbial type consumes a smaller type. Some types of protozoa can ingest and digest bacterial cells. Protozoa regulate certain bacterial populations in the rumen of sheep and cattle.

Physiological influences on the host Small bowel motility is faster in conventional than in germfree animals. Slowing the flow of bowel contents by means of drugs increases susceptibility to intestinal infection.

Priming immunological defence mechanisms Tissues close to body sites harbouring a normal microflora tend to be rich in cells involved in nonspecific and specific immunological mechanisms. Germfree animals have poorly developed reticuloendothelial tissues, and less gamma globulin (antibody) in their blood compared with conventional animals. Conventional animals are said to have a larger stock of immunocompetent cells than germfree animals.

2. Short chain fatty acids (particularly butyric acid) produced by the normal microflora of the large bowel inhibit the multiplication of salmonella under appropriate conditions of pH (undissociated acids) and E_h (anaerobic).
3. Mucosa-associated members of the normal microflora of the large bowel, in synergy with immunological mechanisms, prevent invasion/ destruction of tissue in the large bowel.
4. Small numbers of salmonella cells entering the tissues of the host are quickly killed, the immunological tissues having been nonspecifically stimulated by antigens associated with certain members of the normal microflora.

The protective effect of the normal microflora, however, can be overcome by the entry of large numbers of a pathogen into the ecosystem. In other words, the regulatory mechanisms operating in the ecosystem cannot be relied on to provide us with complete protection from disease. The specific (antibodies, cellular aspects of immunity) and nonspecific (secretions, lysozyme, epithelial integrity, polymorphonuclear leucocytes) factors of host origin constitute other barriers to infection which a pathogen must face. Indeed, it is difficult to consider microbial interference in isolation from host factors since they work together to produce an effect greater than the sum of their individual effects (synergism).

Table 4.6 Applications of microbial interference in human and veterinary medicine

Microbes used as 'probiotic'	Purpose
'Lactic acid bacillus'	Herter, C. (1897) Cultures injected into the small bowel of dogs reduced indican levels in urine.
Lactobacillus bulgaricus	Metchnikoff, E. (1907) Ingestion of cultures of the 'Bulgarian bacillus' reduced the concentration of microbially-derived substances in urine.
Lactobacillus acidophilus	Rettger, L. (1935) 'Acidophilus milk' used to treat cases of constipation, chronic diarrhoea, colitis, sprue and eczema.
Lactobacillus acidophilus	Chr. Hansens Biosystems. (1980s) Drenches and powders for administration to domestic animals.
Escherichia coli	Nissle, (1916) Commercial preparation named 'mutaflor' for the treatment of intestinal infections and constipation.
Escherichia coli	Duval-Iflah, Y. (1982) Plasmid-free strain that reduces colonization of infant intestine by antibiotic-resistant E. coli.
Staphylococcus aureus	Shinefield, H. (1961) Preventing colonization of infants by pathogenic S. aureus (Chapter 3).
Mixed anaerobes	Nurmi, E. (1973) Inoculation of one-day-old chickens to increase resistance to salmonella infection (Chapter 6).

Ever since antagonistic interactions were first observed between strains of bacteria cultured in the laboratory, the history of microbiology has contained attempts to use living microbes to boost resistance to infectious diseases (Table 4.6). Such studies involved the inoculation of animals, including humans, with a specially chosen microbial strain. Colonization of the appropriate region of the inoculated animal by the microbe would, it was proposed, increase resistance to infectious diseases through improved microbial interference. The use of living microbial preparations to increase host resistance to disease is nowadays allied to the 'probiotic' concept. Derived in the 1960s, the word 'probiotics' originally referred to substances produced by one organism which promoted the growth of another type of organism. The concept has been expanded in recent years to include living microbial preparations which can be used to inoculate animals and which, among other benefits, will promote nonspecific resistance to disease. Probiotics are discussed further in Chapter 6.

FURTHER READING

Clarke, R.T.J. and Bauchop, T. (1977) *Microbial Ecology of the Gut*, Academic Press, London.

Cockerill, F.R., Muller, S.R., Anhalt, J.P. *et al.* (1992) Prevention of infection in critically ill patients by selective decontamination of the digestive tract. *Annals of Internal Medicine* **117**, 545–53.

Drasar, B.S. and Barrow, P.A. (1985) *Intestinal Microbiology*, American Society for Microbiology, Washington, D.C.

Drasar, B.S. and Hill, M.J. (1974) *Human Intestinal Flora*, Academic Press, London.

Gordon, H.A. and Pesti, L. (1971) The gnotobiotic animal as a tool in the study of host microbial relationships. *Bacteriological Reviews* **35**, 390–429.

Hobson, P.N. and Wallace, R.J. (1982) Microbial ecology and activities in the rumen, Parts I and II. *CRC Critical Reviews in Microbiology* **9**, 165–225 and 253–320.

Hungate, R.E. (1966) *The Rumen and its Microbes*, Academic Press, New York.

Luckey, T.D. (1963) *Germfree Life and Gnotobiology*, Academic Press, New York.

Midtvedt, T. (1985) The influence of antibiotics upon microflora-associated characteristics in man and animals, in *Progress in Clinical and Biological Research*, vol. 181, (ed. B.S. Wostmann), Alan R. Liss, New York, pp. 241–44.

Pasteur, L. (1885) Observations relatives a la note precedente de M. Duclaux; par M. Pasteur. *Comptes Rendus Hebdomadaires des Seances de l'Acadamie des Sciences (Paris)* **100**, 68.

Waaij, D. van der (1992) Selective gastrointestinal decontamination. *Epidemiology and Infection* **109**, 315–26.

5 Undesirable company: the role of the normal microflora in disease

The microflora, despite the connotation 'normal', is a collection of microbes whose activities are directed at exploiting their environment so as to permit their cellular replication and continuation of their species. 'Normal', in this case, denotes that the microflora is commonly encountered in clinically healthy humans, and not that the microflora necessarily confers normality on the host. Many members of the microflora have undesirable attributes from the point of view of the host. The role of some members of the normal microflora in both relatively minor (e.g. body odour) and major (e.g. periodontitis, urinary tract infection) conditions that affect the host has been mentioned in Chapter 1. Among hospital patients, diseases caused by the normal microflora are commonly encountered. These nosocomial (hospital acquired) infections occur because members of the normal microflora, due to surgical or other procedures, coupled with a diminished resistance to infection due to the patient's debilitated state, have the opportunity to escape from the region of the body where they are usually contained and to establish in normally sterile tissues. The normal microflora may also be involved in the causation, or at least exacerbation, of pathologies other than infection in humans and other animal species.

5.1 ANAEROBIC INFECTIONS

The numerically predominant members of the normal microflora are obligately anaerobic bacteria. Although influencing the host in many ways, when confined to their normal body habitats, these bacteria do not damage the tissues of the animal. Confinement is achieved through the presence of an intact epithelium, constitutive body defences (neutrophils, monocytes) and the aerobic nature of tissues. Impairment of any of these factors can result in the translocation of members of the normal microflora to usually sterile tissues where replication of the microbes may occur. These types of infection are termed 'anaerobic infections'. Gas gangrene is a well-known anaerobic infection caused by spore-forming anaerobes, the clostridia

Table 5.1 Infections commonly involving anaerobes

Brain abscess	Extradural or subdural empyema
Chronic otitis media	Dental infections
Pneumonia secondary to obstructive process	Aspiration pneumonia
Lung abscess	Bronchiectasis
Thoracic empyema	Breast abscess
Liver abscess	Pylephlebitis
Peritonitis	Appendicitis
Subphrenic abscess	Other intra-abdominal abscess
Wound infections following bowel surgery or trauma	Puerperal sepsis
Postabortal sepsis	Endometritis
Tubo-ovarian abscess	Perirectal abscess
Gas-forming cellulitis	Gas gangrene

(especially *Clostridium perfringens* type A), whose habitat is the large bowel. The majority of anaerobic infections, however, are due to the activities of nonsporing anaerobes. These anaerobes characteristically produce cavities filled with pus (abscesses).

Anaerobic infections can occur in practically any site in the body (Table 5.1) and can often be associated with a predisposing condition which has permitted members of the normal microflora to escape from the site in which they are usually confined (Table 5.2). Not all of the members of the normal microflora are able to produce anaerobic infections: in general they are anaerobes that are relatively oxygen-tolerant and which presumably have attributes (virulence factors) that enable the bacteria to resist host defence mechanisms. Anaerobes commonly involved in anaerobic infections are listed in Table 5.3. Intra-abdominal abscesses are the best studied types of anaerobic infection. Until relatively recently, they occurred in up to one-third of patients following elective bowel surgery. Surgical procedures involving the colon can permit bowel contents to escape into the abdominal cavity. The administration of antibiotics (e.g. the 5-nitroimidazoles) prior to surgery so that an inhibitory concentration of the drug is present in the patient's tissues greatly reduces the incidence of infection. Intra-abdominal abscesses are common following abdominal trauma due to gunshot or stab wounds in which the bowel is pierced, or after rupture of the appendix in appendicitis. In all of these cases, a large inoculum of facultative and obligate anaerobes enters the abdominal cavity in bowel contents. This material is walled off by phagocytic white cells which rapidly migrate to the site. In the ensuing battle between host cells and bacteria, large amounts of pus are produced. Because of the volatile, odiferous substances produced by the anaerobic metabolism of proteins, pus from anaerobic infections characteristically has a foul odour. Facultatively

Table 5.2 Some examples of conditions predisposing to anaerobic infections

Predisposing condition	Infection
Sepsis following abortion or dental extraction	Brain abscess
Aspiration of oral secretions when unconscious (alcoholism, deep anaesthesia, drug addiction)	Pleuropulmonary infections
Perforation of bowel (appendicitis, malignancy, diverticulosis, surgical)	Intra-abdominal infections including liver abscess
Abortion, malignancy, obstetrical or gynecological surgery	Infections of female genital tract

Table 5.3 Anaerobes commonly isolated from clinical specimens

Bacteroides fragilis, B. thetaiotaomicron
Porphyromonas asaccharolytica
Prevotella intermedia
Fusobacterium nucleatum
Clostridium perfringens
Peptostreptococcus anaerobius, P. asaccharolyticus, P. magnus

anaerobic bacteria (such as *Escherichia coli*) proliferate during the early stages of the infection and this may result in peritonitis. The facultative anaerobes are essential to the production of anaerobic infections, probably because they lower the oxidation–reduction potential of the environment allowing the survival and subsequent multiplication of the obligate anaerobes. They may also be important because they produce essential growth factors utilized by the obligate anaerobes (e.g. vitamin K). Obligate anaerobic bacteria are essential for abscess formation, as has been demonstrated with experimentally induced intra-abdominal abscesses in rats. Anaerobic infections are characteristically mixed infections in which a collection of two to five bacterial species, acting synergistically, may be involved per infection. *Bacteroides fragilis* is the most commonly isolated anaerobe from intra-abdominal abscesses and may have virulence determinants that allow it to establish more readily in tissues than is the case for other bowel bacteria. Little is known of the pathogenesis of anaerobic infections, but the production of a polysaccharide capsule by *B. fragilis* may be important.

Anaerobic infections also occur in animals other than humans. 'Footrot' (foot scald, foot abscess) of sheep is a mixed infection involving three essential bacteria. The microaerophilic bacterium *Actinomyces pyogenes* provides a growth factor for the obligate anaerobe *Fusobacterium necrophorum* during the early stages (epidermal) of the disease. The obligate anaerobe produces a factor which destroys leucocytes which thus helps protect the actinomyces from host defence mechanisms. A keratin-degrading obligate anaerobe, *Dichelobacter nodosus* (*Bacteroides nodosus*), provides a growth factor for *F. necrophorum* later in the disease when the microbes are located in the matrix between the horn of the hoof and the dermis of the foot.

5.2 SEPSIS DUE TO ENTEROBACTERIACEAE

Gram-negative bacilli (e.g. *E. coli*, *Klebsiella* sp., *Proteus* sp., *Enterobacter* sp.) belonging to the family Enterobacteriaceae are commonly responsible for sepsis (infection that produces physiological responses such as changes in body temperature, heart rate, respiratory rate, white cell count) and septic shock (sepsis accompanied by hypotension). The source of these bacteria is usually the patient's own intestinal microflora. While wounds may be contaminated with bacteria from exogenous sources (contaminated bed linen, hands), spread of the bacteria from the intestinal lumen to other tissues can be bloodborne (haematogenous spread). Traumatic or thermal injury to experimental animals is known to increase the permeability of the 'intestinal mucosal barrier' which then permits increased movement (translocation) of bacteria from the lumen into the mucosa and thence to the mesenteric lymph nodes and other parts of the body. The intestinal barrier is imprecisely defined but is thought to be composed of mechanical (intact epithelium) and immunological components (secretory IgA, gut associated lymphoid tissue). The barrier's overall effect is to contain the normal microflora and other gut luminal constituents within the intestinal lumen without relying on an overwhelming inflammatory or immuno-logical response to microbial or other antigens. This is despite the presence of substances in the intestinal contents that are potent inflammatory agents when introduced into body sites other than the intestine. In addition to burns and trauma, predisposing conditions for sepsis include immuno-suppressive treatments, broad-spectrum antibiotic therapy that alters the composition of the normal microflora, anatomic obstruction, intestinal ulceration, malignancies, AIDS and other serious chronic diseases.

5.3 INFLAMMATORY BOWEL DISEASE (IBD)

Defects in the intestinal mucosal barrier are also of significance in the development of chronic inflammatory conditions of the intestinal tract. Ulcerative colitis and Crohn's disease are inflammatory conditions that have a gradual onset, first becoming apparent in early adulthood. The primary mechanism that initiates the disease is unknown, but genetic predisposition to disease is probably involved because familial clustering of cases occurs. It seems unlikely that this clustering is due to a transmissable agent, although a pathogen causing a long latent infection cannot yet be excluded. Indeed, Crohn's disease has many similarities to an infection of sheep due to *Mycobacterium paratuberculosis* (Johne's disease).

The normal microflora of the intestinal tract may be involved in the development of IBD in the following way. Due to as yet unknown factors, the mucosal barrier effect could be diminished so that increased passage of microbial products (endotoxin, peptidoglycan, peptides, lipids) into the intestinal mucosa occurred. These microbial substances are known to attract neutrophils and initiate an inflammatory response and, later, an immune response. Defective regulation of the inflammatory or immune responses could lead to destruction of the patient's own intestinal tissue (autoimmunity) or a continuing inflammatory response to microbial products from the intestinal lumen.

5.4 CONTAMINATED SMALL BOWEL SYNDROME

The contents of the first two-thirds of the healthy small bowel of humans contain only low numbers of bacteria. Microbial numbers are kept at a low level in this region because of the relatively swift flow of contents through the intestine. Under certain circumstances, however, colonization of the upper regions of the small bowel can occur, with pathological results. This situation is referred to as the 'contaminated small bowel syndrome'. The predisposing conditions to small bowel colonization include strictures of the small bowel due to congential abnormalities, tumours, diverticulosis (sac-like structures resulting from ballooning of the intestinal wall), surgical modifications that leave a 'blind' loop through which bowel contents cannot flow, and childhood malnutrition. These conditions result in the formation of areas of stasis in the upper regions of the small bowel. Because of the reduced flow of intestinal contents through these sites, microbes that are normally restricted to the large bowel can colonize the areas of stasis. Obligate and facultative anaerobes are present in these areas, mostly bacteroides, *Escherichia coli* and enterococci. Colonization of the upper parts of the small bowel by these 'faecal-type' microbes can result in malabsorption conditions. Patients suffering from these syndromes lose weight due to poor absorption of carbohydrate and lipid, excrete abnormally large amounts of lipid in their faeces, and suffer from vitamin and protein deficiency as well as water and electrolyte loss. It is likely that most of these problems result from the deconjugation (Figure 5.1) of conjugated bile salt molecules by bacterial bile salt hydrolase in the region of the small bowel where they would normally be involved in the emulsification and

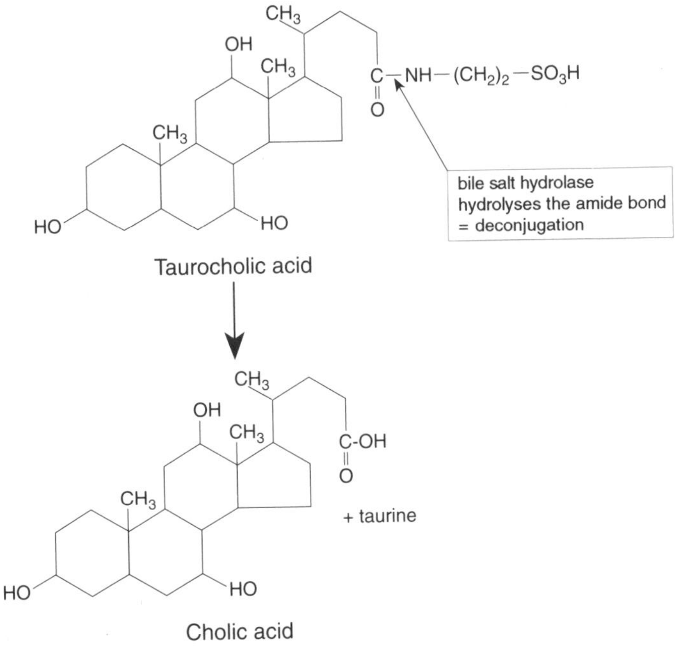

Figure 5.1 Deconjugation of taurocholic acid by bile salt hydrolase.

absorption of lipids (Figure 5.2). The bacteria colonizing the areas of stasis deconjugate the bile acid molecules so that their concentration falls below that necessary for efficient emulsification and micelle formation. Lipid digestion and absorption are therefore impaired, and unabsorbed lipid passes through the remainder of the digestive tract and is excreted in the faeces. The abnormally high concentration of free bile acids resulting from deconjugation damages the epithelium of the upper small bowel. This damage impairs absorption of other nutrients such as carbohydrates and amino acids, and upsets water and electrolyte transport. Vitamin deficiency (e.g. vitamin B_{12}) in patients suffering from malabsorption syndromes is due to utilization or sequestration of the vitamins by the bacteria.

5.5 GROWTH DEPRESSION

In some animal species, notably poultry, pigs and calves, some members of the normal microflora have a growth depressing effect on the host. This effect became evident in the 1950s when it was observed that feeding sub-therapeutic concentrations of antibiotics to animals resulted in an increased weight gain (of the order of 6%). The addition of antibiotics to animal feeds

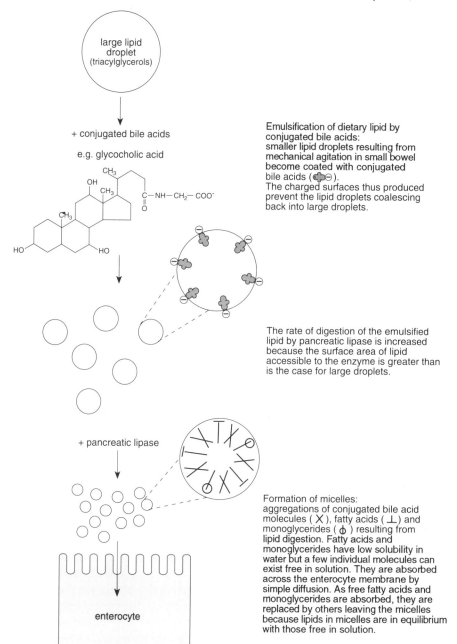

large lipid
droplet
(triacylglycerols)

+ conjugated bile acids

e.g. glycocholic acid

Emulsification of dietary lipid by
conjugated bile acids:
smaller lipid droplets resulting from
mechanical agitation in small bowel
become coated with conjugated
bile acids (⬤⊖).
The charged surfaces thus produced
prevent the lipid droplets coalescing
back into large droplets.

The rate of digestion of the emulsified
lipid by pancreatic lipase is increased
because the surface area of lipid
accessible to the enzyme is greater than
is the case for large droplets.

+ pancreatic lipase

Formation of micelles:
aggregations of conjugated bile acid
molecules (X), fatty acids (⊥) and
monoglycerides (φ) resulting from
lipid digestion. Fatty acids and
monoglycerides have low solubility in
water but a few individual molecules can
exist free in solution. They are absorbed
across the enterocyte membrane by
simple diffusion. As free fatty acids and
monoglycerides are absorbed, they are
replaced by others leaving the micelles
because lipids in micelles are in equilibrium
with those free in solution.

enterocyte

Figure 5.2 The importance of conjugated bile acids in lipid digestion and absorption.

subsequently became widespread, and indeed is now considered essential, in countries where animals are confined at high density in sheds or pens (intensive farming). About half of the total production of antibiotics in the USA is directed to use in farm animals. The specific microbes responsible for growth depression have not been identified except in the case of poultry, where colonization of the small bowel by *Enterococcus hirae* is possibly involved and which may be analogous to the contaminated small bowel syndrome of humans (deconjugation of bile salts by the enterococci). Other explanations that have been advanced to explain growth depression include the following.

1. Ammonia released by faecal bacteria is toxic to the respiratory tract epithelium. Antibiotics reduce microbial activities responsible for ammonia production.
2. The presence of the normal microflora results in a more rapid turnover of intestinal epithelium (by comparison with germfree animals). The host must divert energy from 'growth' to the replacement of epithelial cells. Antibiotics lessen the microbial load, slowing epithelial turnover.
3. The normal microflora competes with the host for nutrients. Antibiotics, by reducing microbial numbers, reduce the extent of the competition.
4. Antibiotics prevent 'subclinical' infections by pathogens.

These explanations are based on limited experimental findings. It is certain, however, that the growth-promoting effect of antibiotics in animal feeds is due to their effect on bacteria: germfree animals do not show a growth response when fed antibiotics.

5.6 THE NORMAL MICROFLORA AS A RESERVOIR OF ANTIBIOTIC RESISTANCE GENES

Antibiotic resistant members of the normal microflora of human subjects have been detected in many studies. The selection and maintenance of antibiotic resistant bacterial strains is probably mainly due to the use and abuse of these drugs in human medicine. Inappropriate prescribing of antibiotics (e.g. for illnesses of nonbacterial aetiology), poor compliance by treated patients (e.g. not completing the prescribed course of antibiotic) and the availability of antibiotics without medical supervision (in some countries) are likely to be responsible. The resistance genes conserved by the normal microflora can be transferred to disease-producing bacteria. These bacteria then constitute a serious medical problem since infections caused by antibiotic-resistant pathogens are more difficult to treat because the first-choice drugs are ineffective. There is thus a delay in initiating an optimal treatment regimen for the patient until a satisfactory antibiotic is found. Second-choice antibiotics are often more expensive and sometimes more toxic to the patient. Antibiotic resistance is particularly serious if the pathogen has multiple resistance due to the microbe harbouring an R plasmid (a plasmid bearing several resistance genes, each of which confers resistance to a different antibiotic). Some of these plasmids encode the ability to transfer themselves from one bacterial cell to another during conjugation. Antibiotic resistance genes are sometimes part of genetic elements called transposons that have the ability to insert themselves into various DNA molecules (bacterial chromosome, plasmids, bacteriophage genomes) within the bacterial cell. Most transposons are transmitted through plasmid transfer but, in the case of some streptococcal transposons, the genetic elements themselves have the ability to initiate conjugative transfer. Whatever the transfer mechanism, certain antibiotic resistance determinants have become widely disseminated among bacterial species. The *tetM* gene encoding tetracycline resistance, for example, has been detected in both Gram-positive and Gram-negative members of the normal microflora: *Peptostreptococcus* species, *Enterococcus faecalis*, *Fusobacterium nucleatum*, *Gardnerella vaginalis*, and *Veillonella parvula*.

The practice of supplementing animal feeds with antibiotics has contributed to the selection of antibiotic-resistant microbes amongst the normal microflora of farm animals which, in turn, poses potential health problems for human beings. Pathogens that infect both farm animals and humans could acquire antibiotic resistance genes from members of the normal microflora and be transmitted to human subjects. Evidence that antibiotic resistances have increased in frequency in pathogens able to infect both humans and other species of mammals is provided by members of the genus *Salmonella*. Isolations of *Salmonella typhimurium* from calves in England during the 1960s were increasingly found to belong to phage type 29. The majority of these isolates, by 1964, were resistant to aminoglycosides, sulphonamides, tetracycline, ampicillin, furazolidone and chloramphenicol. The increased frequency of isolation of this phage type was related to the way in which young, antibiotic-fed, calves were sold to dealers for later resale to other farmers. Mixing of calves from several farms occurred so that there was ample opportunity for transfer of the pathogen to large numbers of animals which were subsequently redistributed to purchasers throughout England. That antibiotic-resistant *Salmonella* can be transmitted from farm animals to humans has been demonstrated in the USA. Eighty seven percent of *Salmonella newport* isolates from human subjects in California in 1985, for example, were resistant to chloramphenicol. All of the isolates were of the same strain as evidenced by the presence of a single plasmid that was characterized by restriction endonuclease analysis. The same strain of salmonella was detected in hamburger meat purchased by the patients, at the abattoir where the cows from which the meat came had been slaughtered, and on the farms where the cows had originated. Chloramphenicol had been administered to animals on the farms. In studies utilizing genetically marked strains of *Escherichia coli*, transfer of anti-biotic-resistant bacteria between farm animals of various species, between farm and feral animals, and between farm animals and human subjects involved in the husbandry of the animals has also been demonstrated.

The precise conditions under which transfer of antibiotic resistance genes among members of the normal microflora occurs has not been deeply studied, but the ratio of potential donor bacterial cells to potential recipient cells has been identified as an important factor in the intestinal tract. The presence of antibiotics in the habitat may be important in some instances for maintenance of R plasmids after transfer, and for induction of con-jugative transfer of resistance determinants in others. Evidence from studies of both the skin and intestinal microfloras has shown that the presence of antibiotic selection pressure is not always essential for transfer and maintenance of antibiotic resistance genes. The presence of antibiotics does, however, result in antibiotic resistant strains becoming more numerous in the habitat.

5.7 CANCER OF THE COLON

The frequency of cancer of the colon varies geographically (Table 5.4). The observed differences in occurrence of the disease between countries are not due solely to genetic (racial) differences because migrants from a 'low' incidence country settling in a 'high' incidence country show a changed (increased) susceptibility to the disease once they assume the cultural habits of their adopted country. Comparison of statistics relating to cultural fac-tors between high and low incidence countries suggests that dietary habits, especially meat (fat and protein) consumption, are linked to the develop-ment of the disease. The inhabitants of high incidence countries consume more meat in their diets than do those living in low incidence countries. In some instances, fibre (plant structural material) is consumed in greater amounts by the residents of low incidence countries. Comparison of current statistics may be misleading: cancer of the colon is a disease of those over 50 years of age and probably takes many years to develop. Lifestyles followed 30 or more years ago are possibly more relevant to understanding the causation of cancer of the colon. Despite the less complete statistics available from previous decades, the relationship between the amount of meat in the diet and cancer of the colon seems to be retained. Thus dietary differences have been invoked in the explanation of the causation of cancer of the colon. Dietary components and host secretions can be metabolized by members of the normal microflora of the large bowel so it has been pro-posed that cancer of the colon may be induced by chemicals produced in the large bowel by the metabolic activities of microbes. Mutagens (car-cinogens are usually also mutagenic) are more commonly present in the

Table 5.4 Geographical incidence of colon cancer

Very low rates (< 5/100 000)
  Ibadan, Nigeria
  Cali, Colombia
  Bombay, India
  Non-Jews of Israel
Low rates (5–10/100 000)
  Miyagi, Japan
  Puerto Rico
  Cuba
  Bulawayo, Zimbabwe
  Finland
  Israel-born, African and Asian-born of Israel
Medium rates (10–19/100 000)
  South metropolitan England
  Sweden
  Denmark
  Norway
  Hawaiians and Filipinos of Hawaii
  Chinese of Singapore
  European and American-born of Israel
High rates (20 or more/100 000)
  British Columbia, Canada
  Non-Maori of New Zealand
  Caucasian, Black and Chinese of San Francisco, USA
  Caucasian, Chinese and Japanese of Hawaii

Colon cancer incidence in selected countries (subnational registry in some cases). Rates per 100 000 population; age-standardized on world population. (Source: Foster, F.H. (1977) *Proceedings of Medical Research Council Cancer Workshop.* Medical Research Council of New Zealand.)

faeces of the inhabitants of high incidence countries than of low incidence countries, and isolates belonging to the genus *Bacteroides* have been isolated which can synthesize a mutagen (a fecapentaene) under laboratory conditions. It remains to be shown, however, whether this class of mutagen is directly relevant to the causation of colon cancer in humans. Bacterial transformation of steroid molecules (bile acids) in the intestinal tract may result in the formation of carcinogenic or co-carcinogenic substances. Cyclopentaphenanthrenes, for example, are carcinogenic molecules that can, in theory, be derived from bile acids. Some intestinal bacteria (e.g. *Clostridium paraputrificum*) are known to be able to carry out at least one of the steps in the sequence by which the bile acid molecule could be converted to a cyclopentaphenanthrene. Diets rich in fat could lead to a greater concentration of such carcinogens in the intestinal tract because more bile (more bile acids) enters the digestive tract in response to increased amounts of fat in the diet. Toxic molecules ingested with, or produced in response to, dietary components might be activated in the intestinal tract due to the metabolism of microbes. This could include dietary molecules (e.g. poly-cyclic hydrocarbons such as benzopyrene resulting from cooking meat at high temperatures) that had already passed through the liver and been detoxified before returning to the digestive tract in an enterohepatic circulation.

There is certainly no shortage of speculation on the ways in which the normal microflora of the intestinal tract might contribute to the causation of cancer of the colon. Solid evidence that they do so is, however, still lacking. Studies to date do not show any consistent changes in the composition of the species comprising the faecal microflora in relation to changes in diet except in the case of *Sarcina ventriculi* which occurs only in the faeces of vegetarians. The lack of more startling differences may reflect inadequacies in the technology presently available for use in such studies. Probably, though, the normal microflora is remarkably constant at the species level because of the reliance of many microbes on host-derived substances, rather than dietary ones, for nutrients. While qualitative and quantitative changes in the normal microflora may be difficult to detect, the induction of microbial enzymes (e.g. β-glucuronidase) in response to changes in diet may well be a more profitable area of investigation in the future.

The amount of vegetable fibre ingested in the diet has been said to influence the causation of cancer of the colon. Burkitt observed a lower incidence of cancer of the colon, among other diseases, in black Africans compared to Europeans. The Africans ingested more fibre in their diet than did Europeans. The frequency of bowel movements was greater in the African population, as was faecal bulk. The retention of water in the large bowel contents by the ingestion of high fibre diets has been suggested to be protective against cancer of the colon: the bulking effect of the fibre would dilute any toxic molecules produced in the intestinal tract. Fibre consumption is, however, about the same in Japan (low incidence), Europe and the USA (high incidence). Similar amounts of fibre, and similar types of fibre, are present in the diets of New Zealand Maori (low incidence) and non-Maori (high incidence). Thus dietary fibre cannot be the complete explanation.

FURTHER READING

Bone, R.C. (1993) Gram-negative sepsis: a dilemma of modern medicine. *Clinical Microbiology Reviews* 6, 57–68.

Finegold, S.M. (1977) *Anaerobic Bacteria in Human Disease*, Academic Press, New York.

Gorbach, S.L. and Goldin, B.R. (1990) The intestinal microflora and the colon cancer connection. *Reviews of Infectious Diseases* 12, S252–61.

Hentges, D.J. (ed.) (1983) *Human Intestinal Microflora in Health and Disease*, Academic Press, New York.

Levy, S.B. (1992) *The Antibiotic Paradox. How Miracle Drugs are Destroying the Miracle*, Plenum Press, New York.

Phillips, S.F., Pemberton, J.H. and Shorter, R.G. (eds) (1991) *The Large Intestine*, Raven Press, New York.

Sherris, J.C. (ed.) (1990) *Medical Microbiology*, 2nd edn, Elsevier, New York.

Speer, B.S., Shoemaker, N.B. and Salyers, A.A. (1992) Bacterial resistance to tetracycline: mechanisms, transfer, and clinical significance. *Clinical Microbiology Reviews* **5**, 387–99.

Spika, J.S., Waterman, S.H., Soo Hoo, G.W. *et al.* (1987) Chloramphenicol-resistant *Salmonella newport* traced through hamburger to dairy farms. *New England Journal of Medicine* **316**, 565–70.

Visek, W.J. (1978) The mode of growth promotion by antibiotics. *Journal of Animal Science* **46**, 1447–69.

Walker, A.R.P. (1976) Colon cancer and diet, with special reference to intakes of fat and fiber. *American Journal of Clinical Nutrition* **29**, 1417–26.

6 Internal renewal: the potential for modification of the normal microflora

Elie Metchnikoff (1845–1916), well known for his pioneering observations and descriptions of phagocytosis, is also noteworthy for his views on the process of ageing. Formulated and couched in the 'natural history' perception of science prevalent in Metchnikoff's time, his books convey an overall message that the normal microflora of the large bowel of humans is detrimental to host welfare. According to Metchnikoff, the large bowel was an asylum for microbes that produced substances that were toxic to the vascular and nervous systems of the host. These toxic substances, absorbed from the intestinal tract and circulating in the bloodstream, contributed to the ageing process. Intestinal microbes were thus identified as the causative agents of 'autointoxication'. The offending microbes were capable of degrading proteins (putrefaction), releasing ammonia, amines and indole which, in appropriate concentrations, were toxic to human tissues. It is true that microbial metabolites that are potentially harmful to tissues are absorbed from the intestinal tract into the blood circulation, but as long as liver function is normal, detoxification of these microbial metabolites occurs so that systemic tissues are not affected. But Metchnikoff inferred that low concentrations of toxins could escape the detoxification mechanisms. His solution for the prevention of autointoxication was radical: surgical removal of the large bowel. A less frightening remedy, however, was to attempt to replace or diminish the number of putrefactive microbes in the intestine by enriching the normal microflora with bacterial populations that ferment carbohydrates to obtain energy and that have little proteolytic activity. Oral administration of cultures of fermentative bacteria would, it was proposed, 'implant' the beneficial microbes in the intestinal tract. Lactic acid-producing bacteria were favoured as fermentative microbes to use for this purpose since it had been observed that the natural fermentation of milk by these organisms prevented the growth of nonacid-tolerant microbes, including proteolytic forms. If a lactic fermentation prevented the putrefaction of milk, would it not have the same effect in the digestive tract if appropriate microbes were used? Eastern European peasants, some of whom were apparently long-lived, consumed fermented dairy products as part of their diet. Thus fermented milk products were introduced to Western Europeans as health-related foods. Although there is conflicting experimental data on the subject, it appears that the lactic acid-producing bacteria (lactic acid bacteria) used in preparing yoghurt (*Streptococcus salivarius* subspecies *thermophilus*; *Lactobacillus delbrueckii* subspecies *bulgaricus*) cannot survive in the human digestive tract. Other species of lactic acid bacteria (Table 6.1), chosen because they are represented among the intestinal microflora, have therefore been used more recently in the preparation of milk products intended to promote health.

Using lactic acid bacteria to produce a fermentative, rather than proteolytic, microflora in the large bowel is, of course, no longer a valid concept. As described in Chapters 1 and 4, the fermentation of carbohydrates of both dietary and host origin, not proteolysis, is the major means by which the members of the normal microflora generate energy in the digestive tract of their host. The concept of implanting 'beneficial' microbes in the digestive tract through administering dietary supplements containing bacterial cultures has, however, been broadened in more recent times to include modification of the microflora as a means of increasing nonspecific resistance to infection or to boost host nutrition. Many commercial products containing living microbes of intestinal origin that are destined for consumption with the aim of improving 'intestinal balance' are available nowadays and are referred to as 'probiotics'. Lactic acid bacteria are

Table 6.1 Lactic acid bacteria used in the preparation of fermented milk products (thermophilic fermentation) which are claimed to promote health

Product	Country	Bacteria
AB-fermented milk	Denmark	*Lactobacillus acidophilus* *Bifidobacterium bifidum*
A-38 fermented milk	Denmark	*L. acidophilus* and mesophilic lactic acid bacteria
Acidophilus milk	USA	*L. acidophilus*
Bioghurt	Germany	*Streptococcus salivarius* ss. *thermophilus*, *L. acidophilus*
Liquid yoghurt	Korea	*Lactobacillus delbrueckii* ss. *bulgaricus*, *L. casei* or *L. helveticus*
Real Active	UK	Yoghurt culture, *B. bifidum*
Miru-Miru	Japan	*L. acidophilus*, *L. casei* *Bifidobacterium breve*
Yakult	Japan	*L. casei*

Table 6.2 Microbes commonly utilized in probiotics

Products for human beings	Products for farm animals
Lactobacillus acidophilus	Lactobacillus acidophilus
L. casei ss. rhamnosus	L. casei
L. casei Shirota strain	L. plantarum
L. delbrueckii ss. bulgaricus	L. delbrueckii ss. bulgaricus
Bifidobacterium adolescentis	Bifidobacterium bifidum
B. bifidum	Bacillus subtilis
B. breve	Streptococcus salivarius ss. thermophilus
B. longum	Pediococcus pentosaceus
B. infantis	Enterococcus faecium
Streptococcus salivarius ss. thermophilus	Saccharomyces cerevisiae
	Aspergillus oryzae
	Torulopsis spp.

popular choices for use in probiotics (Table 6.2), partly because of the historical concept outlined above. Other reasons include (a) certain species of lactic acid bacteria have long been used in the production of foods based on fermented milk (cheese, yoghurt) without harm to the consumer and therefore, as a group, are 'generally regarded as safe' (GRAS); and (b) the large-scale culture of lactic acid bacteria is possible and already developed by the dairy industry. The use of probiotics is not, in theory, limited to the gastrointestinal tract but, apart from staphylococcal interference studies (described in Section 1.2), a few preliminary experiments involving viridans streptococci in the oropharynx, and speculation on the use of lactic acid bacteria as probiotics in the female urogenital tract, most research has centred on the digestive tract.

Members of the genus *Lactobacillus*, many species of which are intestinal inhabitants, have been obvious choices for use in probiotics as they are both lactic acid-producing and intestinal colonizers. The lactobacilli can be accepted as suitable components of probiotics to be administered to pigs and fowl because the bacteria colonize in large numbers the proximal regions of the digestive tract of these animal species, and are present throughout the remainder of the gastrointestinal tract. The use of lactobacilli in probiotics for human use is curious, however, because they are not numerically dominant in the intestinal tract and are absent from the microflora of about 25% of human subjects. Bifidobacteria, however, when present in the human intestinal tract, are about the seventh most numerous bacterial group in faeces. Bifidobacteria are obligate anaerobes that have fermentative metabolism producing major amounts of acetic and lactic acids. The use of these bacteria in increasing numbers of probiotic products may reflect the realization that the bifidobacteria are numerous in the human intestinal tract.

It may be difficult to achieve colonization of the digestive tract of adults by probiotic bacteria even when strains of intestinal origin are utilized. Adult animals are already colonized by an established microflora which will prevent the establishment of the introduced strain, just as it would a pathogen entering the ecosystem. In studies reported so far, it appears that large numbers of microbial cells need to be ingested on a daily basis to achieve persistence of the probiotic strain in the digestive tract. Inoculation of neonatal animals with probiotic bacteria might be more successful since the digestive tract at this stage resembles that of a germfree host. The fate of probiotic strains in the digestive tract has not been studied to any extent due to the lack of methods by which specific strains of bacteria can be detected among the complex microflora of the gastrointestinal tract. Highly specific DNA probes and genetic fingerprinting methods for use with bacteria should enable this necessary information to be generated.

As described in Chapter 2, the colonization of the digestive tract proceeds in a predictable manner until a microflora characteristic of the adult host is attained. The acquisition and maintenance of the microflora is achieved without intentional intervention on the part of humans. The need for the consumption of microbes as dietary supplements must, therefore, be questioned. Why is the introduction into the digestive tract of bacteria contained in capsules or fermented milk products more likely to be satisfactory than natural exposure to these bacteria, considering that they have the same source in nature? The bacteria contained in probiotics might have special properties that would enhance the health of the host, but this must be considered unlikely because probiotic strains are currently chosen largely on the basis of their suitability for industrial-scale cultivation and survival ability during storage. Most, if not all, of the characteristics of these strains have been determined under *in vitro* conditions, and little is known of their activities *in vivo*.

The use of probiotic preparations on the advice of a medical practitioner is generally limited to the consumption of yoghurt after treatment with an antibiotic administered by the oral route. Oral antibiotic administration can result in alterations to the normal microflora of human subjects with associated changes in bowel habits. General practitioners sometimes advise patients to consume yoghurts to 'restore the microbial balance' in their intestinal tract. According to anecdotal reports, the diarrhoea accompanying or following antibiotic usage ceases about the same time as the consumption of fermented milk occurs. Since the bacteria utilized commonly in the manufacture of yoghurt apparently do not colonize the gastrointestinal tract, any amelioration of antibiotic-associated diarrhoea cannot be due to a reconstitution of the microflora with the lactic acid bacteria. Coincidental resolution of the problem as the microflora reconstitutes naturally, a constipating effect of the milk product, or a placebo effect are possible, alternative explanations.

Modern intensive farming methods provide conditions under which young animals may have little, if any, contact with adult animals and are raised under clean conditions relative to the traditional farmyard. The opportunities for the acquisition of an intestinal microflora may be decreased under these circumstances. The acquisition of a digestive tract microflora that confers a degree of nonspecific resistance to some infectious diseases (Section 4.3) is important because young animals have a high susceptibility to infection by intestinal pathogens. Probiotics may therefore be useful in guaranteeing an appropriate intestinal microflora early in life. Studies relating to the nonspecific resistance of chickens to *Salmonella* colonization, however, show that the resistance is mediated by a collection of at least 30 bacterial species (mostly obligate anaerobes) and not by single bacterial types. Stressful conditions are known to alter the microflora, including a decrease in the lactobacillus population, of the digestive tract of farm and experimental animals. It is unlikely, though, that the administration of lactic acid bacteria in times of stress could lead to recolonization of the tract: if conditions in the intestine were unsuitable for the replication of indigenous lactobacilli, they would be equally unsuitable for colonization by the probiotic strain to occur.

In addition to contributing to a 'balanced intestinal microflora', it has been proposed that probiotics could enhance resistance to infectious disease by stimulating components of the immunological system. The non-specifically-stimulated immunological system would then be ready to quickly destroy invading pathogens. It has been demonstrated experimentally that the cell wall components of lactic acid-producing bacteria stimulate the immunological system. Cell wall material from *Bifidobacterium longum, Bifidobacterium thermophilum, Lactobacillus casei, Lactobacillus delbrueckii* ss. *bulgaricus* and *Lactobacillus plantarum* has been reported to influence immunological phenomena in various animal species. The major effect exerted by the bacterial cell wall material seems to be upon macrophages (mononuclear cells that are free in tissues or attached to walls of blood sinuses). Macrophages can ingest and destroy microbial cells and other 'foreign' cells or materials, or damaged body components, and contribute greatly to the inflammatory response through the release of interleukin-1 and tissue necrosis factor. These substances enhance neutrophil activities. Like neutrophils, the cytoplasm of macrophages contains membrane-bounded sacs (lysosomes) that contain hydrolytic enzymes, cationic antibacterial proteins and molecules that catalyse the formation of oxygen radicals that are toxic to microbial cells. Phagocytozed bacteria are contained within vacuoles (phagosomes) in the macrophage cytoplasm. Lysosomes migrate to the phagosome and fuse with it, liberating their contents. The bactericidal and digestive processes of the lysosomal contents kill and destroy the bacteria. Unlike neutrophils, macrophages continue to differentiate after they have left the bone marrow and, under conditions of appropriate stimulation, become 'activated'. Macrophages that have been activated phagocytoze more vigorously, take up more oxygen and secrete larger quantities of hydrolytic enzymes. In general, activated macrophages are better able to kill microbial cells. *Lactobacillus casei* material, for example, activates macrophages so that they have increased oxygen radical production (e.g. of superoxide anions) and increased lysosomal enzyme activity, have increased phagocytic ability and are induced to release colony-stimulating factor (causing stimulation of proliferation and differentiation of granulocytes and macrophages). As well as their role in nonspecific resistance, macrophages participate in the induction of specific immune responses by stimulating the development of lymphocytes.

The stimulation of the immunological system by lactic acid bacteria must be considered in terms of its biological significance. In experimental studies, has the immunostimulatory effect of the probiotic been produced using conditions that resemble those in real life? Does the immunostimulation produce long-term resistance to infectious disease, reduce the period of illness or decrease mortality? In general, the studies reported so far in the scientific literature do not meet these criteria. For example, large amounts of bacterial cells or cell wall material have been used to inoculate experimental animals by parenteral (other than by mouth) routes. These conditions do not resemble the consumption of fermented milks or probiotic products.

Nonspecific stimulation of the immunological mechanisms of the body should be approached with caution since, in some subjects, components of the immune system that are activated and respond to the bacterial substances might also react with host components producing an autoimmune disease. Mycobacteria, for example, are potent activators of macrophages and are currently being investigated for their involvement in the causation of rheumatoid arthritis in humans. Cell wall material from *Bifidobacterium breve*, *Bifidobacterium adolescentis*, *Eubacterium aerofaciens* and *L. casei* produces moderate to severe arthritis in rats inoculated intraperitoneally with the bacterial components.

6.1 PROMISING DEVELOPMENTS IN PROBIOTIC RESEARCH

Probiotic products that are presently available can be criticized for their lack of a sound scientific basis, but future research may rectify this situation. Probiotics that provide predictable outcomes able to be measured scientifically are required; the same criteria used in developing pharmaceutical products. Probiotic research in three areas currently appears promising.

6.1.1 Fermented milk products for lactose-intolerant subjects

The disaccharide lactose, which constitutes about 5% of cows' milk, is hydrolysed in the small bowel of children, with the release of glucose and galactose. The hydrolysis is catalysed by the lactase (a β-galactosidase) associated with the brush border of enterocytes lining the intestinal tract. Glucose and galactose can be absorbed and metabolized. Congenital deficiency in lactase is extremely rare, but late-onset hypolactasia developing during childhood is common in many regions of the world including among Asian, Arab and African populations (80–100% of individuals). The condition also occurs in caucasians, but at a lower incidence (25%). Deficiency in the production of lactase in the intestinal tract results in the passage of dietary lactose in an unaltered state to the lower small bowel and the large bowel. Some members of the normal microflora inhabiting these regions produce β-galactosidases. The products of lactose hydrolysis are fermented by the bacteria with the production of gases (including hydrogen) and organic acids, giving rise to the symptoms of abdominal discomfort and osmotic diarrhoea experienced by lactose intolerant individuals.

Lactose intolerance is defined clinically as the occurrence of gastrointestinal symptoms after administration of a single test dose of 50 g of lactose in aqueous solution. More specifically, lactose-intolerant individuals have a breath hydrogen concentration over 20 ppm during an 8 hour period following ingestion of 20 g of lactose. This breath test provides a standard, internationally recognized, noninvasive clinical assay by which lactose intolerance can be diagnosed, and the effects of fermented milk products on lactose-intolerant individuals can be evaluated under conditions that exclude a placebo effect. Lactose-intolerant individuals can tolerate fermented milk (yoghurt) better than nonfermented milk containing the same amount of lactose. This observation is matched by the results of hydrogen breath tests. It appears that, although the lactic acid bacteria themselves cannot replicate in the small bowel, β-galactosidases produced by *S. salivarius* subspecies *thermophilus* and *L. delbrueckii* ss. *bulgaricus* during the production of the yoghurt can pass through the stomach environment without denaturing (the enzyme is intracellular). The β-galactosidases are inactive at the acid pH of the final yoghurt product and do not hydrolyse lactose until the digesta has passed well down the small bowel. By this time, the pH of the digesta has reached a pH near neutrality which is the optimal pH for the β-galactosidases produced by the streptococci and the lactobacilli. Additionally, the bacterial cells may lyse as they pass through the small bowel. It must be assumed that sufficient amounts of glucose and galactose, resulting from lactose hydrolysis, are absorbed from the intestine before the food bolus reaches sites of microbial colonization. Thus microbial fermentation of the carbohydrates does not occur to an extent that would produce the symptoms of lactose intolerance.

6.1.2 Influence of lactobacilli on enzyme activities in the large bowel

Cancer of the large bowel is a major cause of mortality in countries such as New Zealand, the UK and the USA (Section 5.7). Although dietary composition does not affect markedly the species composition of the intestinal microflora, diet does influence enzyme activities, of microbial origin, in the digesta. Many factors are likely to be involved in the development of cancer of the colon. Bacterial enzymes may be one such factor, since they could form or release carcinogens in the large bowel. Azoreductases have received attention, for example, because these enzymes, produced by a wide range of bacterial species comprising the normal microflora, catalyse the cleavage of azo bonds in dyes used in the food industry as colouring agents. These azoreductases have the potential to mediate the formation of toxic amines (some of which are carcinogens) in the intestinal ecosystem.

Beta-glucuronidase catalyses the cleavage of glucuronic acid molecules from glucuronides entering the intestinal tract in bile. Produced mainly by *Escherichia coli* and obligate anaerobes such as bacteroides and clostridia, β-glucuronidase activity could be significant in terms of the reactivation of potentially toxic molecules, including carcinogens or co-carcinogens, that had been detoxified by the formation of glucuronides in the liver. The involvement of azoreductase and β-glucuronidase activities in the causation of cancer of the colon is speculative. However, efforts at modifying the amount of their activity by the use of a lactobacillus strain administered to experimental subjects in milk have been made. In a study by Goldin, Gorbach and colleagues, β-glucuronidase and azoreductase (which is also a 'nitroreductase') activities were monitored in the faeces of human subjects during successive periods in which they consumed their usual diet, or supplemented their diet with low fat milk or with milk containing *L. acidophilus* (about 10^9 lactobacilli per ml). During the period when the diet was supplemented with *L. acidophilus*, a two to four-fold reduction in the activity of the enzymes was recorded.

Unfortunately, this study did not evaluate the lactobacillus population of the intestinal tract during the course of the experiment. It is not known, for example, whether the *L. acidophilus* strain administered to the subjects in milk survived, let alone metabolized, in the digestive tract. Investigations using experimental animals, however, have demonstrated that colonization of the digestive tract by lactobacilli influences azoreductase and β-glucuronidase activities. These experiments used animals that were colonized by a complex gastrointestinal microflora functionally equivalent to that of conventional mice, as judged by the determination of 26 microflora-associated characteristics, but were not host to lactobacilli (lactobacillus-free mice). Maintained in isolators by gnotobiological methods, the colony of mice provided a microbiologically-constant system with which to determine the influences of lactobacilli when resident in the digestive tract of the host. Comparison of the biochemistry of the large bowel content of lactobacillus-free mice with that of animals with an identical microflora but to which had been added lactobacillus strains derived from conventional mice, showed that the lactobacilli influenced enzymatic activities. Azoreductase activity was 31% lower in animals colonized by lactobacilli compared to those that were lactobacillus-free. Male lactobacillus-free mice had 52% more β-glucuronidase activity in the caecum than female mice. The enzyme activity was reduced to that observed in females when lactobacilli colonized the digestive tract of the male mice. The mechanisms by which lactobacilli mediate these phenomena and the significance of the altered biochemistry to the animal host are presently unknown. Collectively, the results from experiments with human subjects and with mice indicate that further research in this area is justified.

6.1.3 The Nurmi concept

Concern over the incidence of *Salmonella* colonization in poultry flocks in Finland during the 1970s led to the observations by Nurmi and colleagues that the intestinal tract of day-old chicks could be rendered more resistant to colonization by *Salmonella* if the birds were inoculated orally, immediately after hatching, with a homogenate prepared from the faeces or intestinal contents of an adult fowl. The rationale for this procedure was that since incubator-hatched chicks did not have contact with adult poultry, they lacked exposure to an appropriate source of normal microflora members. The acquisition of their intestinal microflora would be, at the very least, delayed and microbial interference phenomena would not operate in the chicken intestine. The chicks were thus susceptible to colonization by *Salmonella* commonly present as contaminants of poultry feed. Inoculation of the chicks with an intestinal inoculum resulted in the acquisition of an appropriate microflora and this, in turn, led to an increased resistance to *Salmonella* colonization. The interference phenomenon was found to be specific: inoculation with intestinal contents of poultry origin was effective; inoculation with rumen contents or a homogenate of horse faeces was not. Using intestinal homogenates from fowl to inoculate chicks risked exposing the young birds to viral pathogens (e.g. the agent of Marek's disease). Mixed (often undefined) cultures of intestinal microbes from adult fowl have therefore been developed for inoculation of day-old chicks. Composed largely of anaerobic bacterial types, the mixtures contain at least 30 species. They are difficult to prepare in a standardized manner but appear to be effective under field conditions, especially if the preparations are used to spray the newly hatched birds rather than relying on the chicks' drinking inoculated water.

6.2 A NEW GENERATION OF PROBIOTICS

Advances in molecular genetics permit the derivation by genetic modification of bacterial strains that have novel attributes. Although knowledge of the genetics and molecular biology of normal microflora members other than *E. coli* is scarce, the potential exists for the derivation of probiotic strains in which genes encoding undesirable attributes have been inactivated, and genes encoding desirable characteristics introduced. The following types of probiotics might be developed through the use of recombinant DNA technology.

6.2.1 Vaccines

A member of the normal microflora could be genetically modified so that its cells synthesized an immunogen characteristic of a particular pathogen. Colonization of the gastrointestinal tract, say, with such a strain could result in a continuous exposure of the intestinal mucosa to the immunogen so that secretory IgA antibodies would be synthesized by the host animal. This could result in immunity, at mucosal surfaces, of the host to the pathogen. It is fortunate that stimulation of the immunological tissues associated with the intestinal tract leads to the presence of specific antibodies in the secretions of not just the intestinal tract, but also on the mucosa of other body sites. Thus, potentially, immunity to intestinal, respiratory or genital tract pathogens could be produced using a suitably modified intestinal microbe. The genetically modified microbe might be transmitted from mother to offspring, so that a self-perpetuating system of immunization would be produced. Genetic modification of bacteria or viruses so that they synthesize a variety of immunogens (multivalent vaccines) has already been applied to avirulent *Salmonella* strains and to vaccinia virus, but the use of members of the normal microflora may be aesthetically more pleasing to the general public and to health regulatory agencies.

6.2.2 Pharmacology

Biotechnological research has led to the derivation of genetically-modified bacterial and yeast strains that can be cultivated in large volumes under industrial conditions. The modified organisms synthesize large amounts of substances of medical and industrial significance. These substances are subsequently harvested and purified from the cultures. It may be possible to derive strains of gastrointestinal or other microbes that similarly synthesize novel substances following genetic modification, but that can be used to deliver molecules with biological activity to particular regions of the intestinal tract without the need to industrially purify the beneficial substance in question. For example, a microbial strain whose cells lysed under specific conditions, either because of biochemical conditions prevailing in a particular part of the digestive tract or because of triggering of a lytic bacteriophage infection (as has been engineered in *Lactococcus lactis*), might be derived. Incorporated into foods, the microbial cells would then deliver a product to a specific level of the intestinal tract.

6.2.3 Nutrition

The contribution of the rumen microflora to the nutritional well-being of ruminants has been described in Chapter 4. Probiotics containing microbial strains that would permit the ruminant to more fully utilize the energy contained within its diet could be derived and introduced into the rumen. The derivation of microbes that produced increased amounts of cellulolytic enzymes, for example, would be of interest because only about 60% of cellulose entering the rumen is degraded. Other possibilities include the derivation of microbial strains that synthesize large quantities of amino acids (lysine, methionine, threonine) that are growth-limiting for the ruminant, or strains that have enhanced ability to convert lactic acid to propionic acid. This would prevent lactic acidosis (abnormal increase in the acidity of the blood due to absorption of lactic acid from the digestive tract) in the host with a radical change in diet (e.g. from high fibre to high grain diets).

Genetically modified microbial strains could also be useful in the inactivation of toxic molecules liberated from the diet during the digestion of food in the gastrointestinal tract. *Leucaena leucocephala* is a tropical leguminous shrub that produces high yields of protein-rich forage in tropical and subtropical regions. Diets rich in *Leucaena* have been associated with ruminant health problems in northern Australia because of a breakdown product of the amino acid mimosine present in plant material. Autolytic or microbial degradation of mimosine produces a potent goitrogen, 3-hydroxy-4(1H) pyridone (DHP). Ruminants ingesting a *Leucaena* diet in Hawaii or Indonesia, however, do not exhibit the symptoms of hypothyroidism seen in Australian ruminants and excrete very little DHP in their urine. It is likely that Hawaiian and Indonesian ruminants harbour rumen microbes capable of detoxifying DHP, since inoculation of Australian ruminants with rumen fluid from Indonesian animals permits the ingestion of a *Leucaena* diet without toxicity problems. These observations raise the possibility of manipulating the rumen microflora of an animal so that it is protected from the toxic substances present in potentially useful pasture plants introduced from another country. The genetic modification

of rumen bacteria so that they can inactivate molecules toxic to the host could be part of this manipulation.

FURTHER READING

Cardenas, L. and Clements, J.D. (1992) Oral immunization using live attenuated *Salmonella* spp. as carriers of foreign antigens. *Clinical Microbiology Reviews* 5, 328–42.

Fuller, R. (ed.) (1992) *Probiotics. The Scientific Basis*, Chapman & Hall, London.

Goldin, B.R., Swenson, L., Dwyer, J., Sexton, M. and Gorbach, S.L. (1980) Effect of diet and *Lactobacillus acidophilus* supplements on human fecal bacterial enzymes. *Journal of the National Cancer Institute* 64, 255–61.

Gracey, M. (1991) Sugar intolerance, in *Diarrhea* (ed. M. Gracey), CRC Press, Boca Raton, pp. 283–97.

Jones, R.J. and Lowry, J.B. (1984) Australian goats detoxify the goitrogen 3-hydroxy-4(1H) pyridone (DHP) after rumen infusion from an Indonesian goat. *Experientia* 40, 1435–6.

McConnell, M.A. and Tannock, G.W. (1991) Lactobacilli and azoreductase activity in the murine cecum. *Applied and Environmental Microbiology* 57, 3664–5.

McConnell, M.A. and Tannock, G.W. (1993) A note on lactobacilli and β-glucuronidase activity in the intestinal contents of mice. *Journal of Applied Bacteriology* 74, 649–51.

Metchnikoff, E. (1907) *The Prolongation of Life. Optimistic Studies*, William Heinemann, London.

Sanders, M.E. (1993) Effect of consumption of lactic cultures on human health. *Advances in Food and Nutrition Research* 37, 67–130.

Teather, R.M. (1985) Application of gene manipulation to rumen microflora. *Canadian Journal of Animal Science* 65, 563–74.

Index

Page numbers in **bold** refer to figures and page numbers in *italics* refer to tables.